Fractured Forest, Quartzite City

Thank you for choosing a SAGE product!
If you have any comment, observation or feedback,
I would like to personally hear from you.

Please write to me at **contactceo@sagepub.in**

Vivek Mehra, Managing Director and CEO, SAGE India.

Bulk Sales

SAGE India offers special discounts
for purchase of books in bulk.
We also make available special imprints
and excerpts from our books on demand.

For orders and enquiries, write to us at

Marketing Department
SAGE Publications India Pvt Ltd
B1/I-1, Mohan Cooperative Industrial Area
Mathura Road, Post Bag 7
New Delhi 110044, India

E-mail us at **marketing@sagepub.in**

Subscribe to our mailing list
Write to **marketing@sagepub.in**

This book is also available as an e-book.

Fractured Forest, Quartzite City

A History of Delhi and Its Ridge

Thomas Crowley

Illustrations by deepani seth

𝟓 YODAPRESS | ⑤SAGE | seleCT

Los Angeles I London I New Delhi
Singapore I Washington DC I Melbourne

First published in 2020 by

SAGE Publications India Pvt Ltd
B1/I-1 Mohan Cooperative Industrial Area
Mathura Road, New Delhi 110 044, India
www.sagepub.in

YODA Press
79, Gulmohar Enclave
New Delhi 110049
www.yodapress.co.in

SAGE Publications Inc
2455 Teller Road
Thousand Oaks, California 91320, USA

SAGE Publications Ltd
1 Oliver's Yard, 55 City Road
London EC1Y 1SP, United Kingdom

SAGE Publications Asia-Pacific Pte Ltd
18 Cross Street #10-10/11/12
China Square Central
Singapore 048423

Published by Vivek Mehra for SAGE Publications India Pvt. Ltd. Typeset in 9.5/13 pt Georgia by Fidus Design Pvt. Ltd, Chandigarh.

Library of Congress Control Number: 2020944535

ISBN: 978-93-5388-554-0 (PB)

SAGE YODA Team: Amrita Dutta, Sandhya Gola, Arpita Das and Ishita Gupta

CONTENTS

Acknowledgments   vii

Seeds: Introduction   1

Chapter 1. Stones: Shifting Geologies of the Ridge   21

Chapter 2. Soil: Mobile Ecologies, Hybrid Histories   61

Chapter 3. State: Warfare, Pageantry, Politics   121

Chapter 4. Surplus: Production, Consumption, Speculation   209

Chapter 5. Spirits: Transcendence, Sacred and Secular   271

Notes   324
Bibliography   337
About the Author   351

ACKNOWLEDGMENTS

I first encountered Delhi's Ridge a few weeks after moving to the city in 2010, when a friend suggested I visit the "monkey park" near my North Delhi flat. It has been a decade between my introduction to the Ridge and the publication of this book. During these years, as the scope of the book expanded in fits and starts, I ventured into many areas quite far from my limited realms of expertise, attempting to synthesize large amounts of primary and secondary literature on the city and its ridge. In this attempt, I've gone out on many limbs, and it seems inevitable that some of them may support my weight better than others. Those branches that stand no doubt do so because of the strength of those supporting me. Their intellectual generosity, patience and support has made this book possible. (Naturally, any breaking branches and falling limbs can be blamed only on my own overreaches.)

P.T. George at Intercultural Resources (ICR) in Delhi made all of this possible by hiring me as a researcher and encouraging my work on the Ridge. I first began to sketch out the current book during two stints in the City as Studio residency organized by Sarai at the Centre for the Study of Developing Societies (CSDS). A large chunk of the manuscript was written while I was a Social Science Fellow at the Akademie Schloss Solitude in Stuttgart, Germany. These institutional and financial supports were absolutely vital for the creation of this book.

From my time at Sarai, I made lasting friendships that also transformed my work on the Ridge. For a time, this took the form of a possibly fictional but extremely generative artist collective, "This is Not Us". TINU, this book is for you.

Ujjwal Utkarsh, a fellow-traveler both figuratively and literally, read through early writing attempts and encouraged me to find a voice for the text. Rashmi Munikempanna helped me think through the political stakes of writing about the Ridge. Agat Sharma pushed me to think artistically about the Ridge, and gave me the invaluable opportunity to incorporate my research into creative workshops at Pearl Academy in Delhi and Noida.

Deepani Seth, who created the indispensable illustrations for this book, has seen this project through from start to finish. Our collaboration started with planning a game in the Ridge for Sarai (thanks to Gabe Smedresman, my game guru, for inspiration on this!) and then expanded to all corners of the Ridge and to far-flung markets, chai stalls, restaurants and parks in Delhi. From working on illustrated articles together (some of which found publishers, some of which remain in publishing purgatory) to debating all things Delhi to conducting workshops together in Noida and Stuttgart, deepani has been there with me as a research collaborator, critic, discussant and dear friend.

In Delhi, a series of flatmates not only made the city a home for me, they also put up with—sometimes even encouraged—my obsession with the Ridge. So a big thank you to Samuel Buchoul, Tanveer Kaur, Nandini Sarkar, Mehrnoush Rezaie, Anjali Pathirana, Mithu Biswas and Rajeeb Kari. Thanks also to Vidhya Raveendranathan for early reading suggestions and pushing me to think about labor, caste and nature in the city. Once I began writing the book, Nehru Memorial Museum & Library (Teen Murti) became my home away from home. The informal community of research scholars and friends there (and the nearby Mysore Cafe) made it possible to write this book. Thank you to Snigdha Kumar, Aban Raza and Amita Rana for the many talks over lunch and far-too-weak chai, and for friendship and support that extended well beyond the walls of Teen Murti.

I am extremely grateful to the scholars, activists and civil society leaders in Delhi who took the time to meet with me and share their encyclopedic knowledge of the city and the Ridge, especially when my research was still quite green (naive, not eco-friendly!). I benefited greatly from the insights of Ravi Agarwal, Aastha Chauhan, Narayani Gupta, Sohail Hashmi, Revathi and Vasant Kamath, Pradip Krishen, Ranjit Lal, Mahesh Rangarajan, Dunu Roy, Kush Sethi, Bhrigupati Singh, Shashank Saini, Anita Soni and Vikram Soni. I am especially grateful to Amita Baviskar for her support for this project over the years and her in-depth, careful reading of the manuscript; her feedback has greatly strengthened and transformed the book.

The staff at the Delhi State Archives helped me access the byzantine files of the erstwhile Deputy Commissioner's Office. Thank you to Janaki Nair at Jawaharlal Nehru University for allowing me to audit her class on "The Modern City", which helped me conceptualize many of the questions I address in this book. Many thanks to the professors and scholars in Delhi and beyond who gave me the opportunity to present my work to a wider audience: Naveen Chander and CACIM's workshop at Sambhaavnaa (and Lalit Batra, Julia Corwin and Tom Cowan for their input there and after); Diya Mehra at South Asian University; Rohit Negi at Ambedkar University Delhi; Amit Ranjan at St. Stephen's College; Renny Thomas at Jesus and Mary College, University of Delhi; and Marie-Hélène Zerah at the CSH-CPR Urban Workshop Series.

This book was nearing completion when I joined the PhD program in geography at Rutgers University. The intellectual community there has given me the strength to finish the book and navigate the world of publishing. Many thanks to my adviser, Asher Ghertner, whose work on Delhi informs the politics of this book; and to his lab group, past and present, for the perfect balance of critical feedback and comradely support: Sangeeta Banerji, Ben Gerlofs, Stuti Govil, Wei-Chieh Hung, Sadaf Javed, Hudson McFann, Priti Narayan, and Devra Waldman.

For helping me conceptualize this project in book form, and giving me insights into the publishing world, thank you to Daniel Levin Becker, Fazal Rashad, Anita Roy, Gavin Steingo and Jonah Walters. At SAGE-Yoda, Arpita Das has been an incisive editor and a generous supporter of the book; and Ishita Gupta has expertly and kindly overseen all the adventures of completing the publishing process across borders.

And finally, I must give profound thanks to my earliest and most exacting editors—Paul Crowley, Resha Crowley and Tara Crowley— and to Meghana Arora, who was my reason for being in Delhi in the first place and who came around to the Ridge in the end.

SEEDS
Introduction

Urban Mirages

Delhi. A chilly mid-November morning in India's capital city. Still groggy from sleep, I step out onto my terrace. Big flakes of snow are falling gracefully in looping, swirling arcs. But something is wrong. The snow is gray and black, and it leaves dark stains on the tiled floor of the terrace. The "snow", I suddenly realize, is sooty, half-burned newspaper, drifting onto my terrace from a nearby garbage fire.

Delhi has not always been this polluted, I am told. Especially not Mehrauli, the area where I live. Mehrauli is Delhi's oldest continuously inhabited neighborhood, home to human settlements since the eleventh century, and perhaps even earlier. In traditional Delhi lore, it is said that seven cities have risen and fallen in Delhi. Mehrauli is the first.

Although Mehrauli lost its role as a center of imperial power in the late thirteenth century, it remained a popular place of residence, in part because of its verdure and natural beauty. It is located on the low, undulating hills of the Aravalli mountain range, and the contours of the land make for dramatic views, with Qutb Minar[1]—the world's tallest brick minaret, and India's most visited tourist site—as the focal point. The neighborhood was particularly prized by the Mughal emperors, whose own capital city was located roughly fifteen kilometers away. The Mughals saw Mehrauli as a pleasant refuge, a peaceful green getaway from the hubbub of city life.

The epitome of this pastoral beauty was a place commonly known as the *jharna* (waterfall). Built in 1700 by a Mughal nobleman, the *jharna* channeled overflow water from a huge man-made lake. The water tumbled down an immense wall before passing through the perforated roof of an elaborate sandstone structure. An admiring observer described the waterfall at night, when lamps illuminated the structure: "it seemed as if someone had set the water on fire, as if there was a heavy shower of melting gold."[2] The water then flowed into an intricate series of canals and pools before reaching a dense mango orchard. In one spot, the water was diverted so that it poured down a smooth flat stone, which was used as a makeshift water slide.

In its heyday, the *jharna*, with its orchards and pools, its slippery stones and its magical roofs, was frequented by the nobility and the common citizen alike. When I visited it in 2012 though, it was eerily quiet and empty. No water flowed through the canals; they were choked with garbage-filled sludge.

Behind the *jharna*, there actually was running water, in the form of a small stream flowing around the base of a low hill. The stream gurgled pleasantly, but its banks were caked with sewage and plastic, the water oddly discolored. Beyond the stream was a small working-class settlement. From a distance, it looked like a village, with tiny picturesque huts on the hillside and trees in the background. But this too was an illusion. The houses were largely made of thin, crumbling brick, with tin roofing and makeshift plastic sheeting to shore up leaky walls.

The sooty snow, the stagnant, putrid stream: Delhi's landscape is mocking its citizens, or shaming them. It presents an ironic inversion of idyllic rural scenes, with seemingly beautiful natural features that reveal themselves to be dangerously toxic. This is the paradox of Delhi: it is the world's most polluted green city.

It has the highest levels of air pollution in the world, at least according to much-publicized 2014 World Health Organization findings.[3] It's not just the air. Delhi has dangerously polluted water bodies, unnerving noise pollution, overflowing landfills and mountains of electronic waste.

And yet Delhi is also remarkably green, especially for a megacity with a population of 17 million, as of the 2011 census (and this is almost certainly an underestimate). The city has an expansive network of small neighborhood parks and big public gardens. The centerpiece of green Delhi is an eighty-square-kilometer zone that the government has set aside as Reserved Forest. This zone is clustered around the Aravalli hills that cradle Mehrauli and several other historic sites in Delhi. It is referred to colloquially as "The Ridge".

Exploring the Ridge

What, exactly, is the Ridge? The book in your hands is an attempt to answer this seemingly simple question. When I started researching the Ridge in 2010, I thought the answer too was simple: it's the city's green lung, its ecological lifeblood, a much-needed forest in the midst of an ever-expanding megacity. The newspaper reports I read gave a clear-cut account of the Ridge's benefits: it purifies Delhi's air, protects it from the hot desert winds of nearby Rajasthan and provides an escape from the madness and speed of urban life. Because of its crucial ecological functions, the Ridge must be preserved "in its pristine glory", to use the words of the government's Master Plan for Delhi.[4]

But this is not the full story. The more I learned about the Ridge, the more complexities I encountered. The seeds I had planted with my initial research began to put down tenacious roots, and to sprout interlocking, entangled branches. The research project was becoming, I feared, an impenetrable thicket of ideas.

A chief confusion was how to define the Ridge. Government reports and environmental groups have emphasized the Ridge's functions as a forest, but the very name of the Ridge suggests that it is, at its core, a geological phenomenon. And indeed, most of the Reserved Forest zone in Delhi corresponds with the Aravalli hills. But there are discrepancies. For instance, there are parts of the Delhi Aravallis that host no trees, let alone a full-fledged forest. Perhaps the most obvious example of this is Mehrauli, which now houses a population of roughly 250,000, packed into multi-story apartments that line labyrinthine streets.

Poring over Ridge-related documents, I began to realize that phrases like "pristine glory" and "the harmony of nature" are serious impediments to understanding the Ridge, since they imply that both its ecology and geology have remained unchanged since time immemorial. The problem with this line of thought is not just the existence of centuries-old settlements like Mehrauli. The problem is also that the forest itself, where it exists, is generally quite new and is dominated by invasive species. What is more, the densest stands of present-day forest are largely located on a pockmarked landscape where, for many decades, quarrying gashed holes in the hills.

The problem, in short, is history. To invoke a timeless balance of nature on the Ridge is to erase its history. But it is precisely in the realm of history that one can discover how the Ridge's geology and ecology have co-evolved, and, even more crucially, how the Ridge and the city of Delhi have shaped each other over the course of hundreds and thousands of years.

My aim, then, is to recover the lost history of the Ridge, and, in the process, to tell a story of Delhi that puts its environment front and center. While there has been no dearth of writing on Delhi, the city's chroniclers often ignore its ecological features. But Delhi looks different when viewed from the heights of its hills, or from the depths of its old mining pits, or from the thickets of its newly grown woodlands.

Every city depends on its geological and ecological foundations, but Delhi's relationship with its Ridge has been particularly long, complex and fraught. Because of the unique breadth and depth of Delhi's history, the tale of the Ridge is one that resonates far beyond the boundaries of India's capital. At various points, Delhi has been a crucial hub of politics, warfare, trade and religious expansion on a regional and even global level. The Ridge offers a crucial vantage point for viewing these historical and geographical interconnections.

A thriving city with millennia of history, Delhi has long attracted people from all over the world, myself included. In an age when refugees are demonized, when xenophobia is on the rise, the history of the Delhi Ridge offers a lesson about the value, and indeed the absolute necessity, of migrants for building a vibrant metropolis. It is an oft-noted fact that Delhi is a city of migrants. This is largely a reference to the flood of refugees who came to the city in 1947 during the trauma of Partition and gave the city its post-Independence flavor. But the proliferation of migrants to Delhi, and specifically to the Ridge, has a much longer history—even into paleolithic times, 100,000 years ago, when the first humans (technically, hominids, but that's a story for a later chapter) ventured into north India, continuing their ancestors' journey out of Africa.

The migrants drawn to Delhi have been a motley mix, including warrior yogis, star-crossed lovers, bandit shepherds and Sufi stoners. And that's just the humans. There are also mischievous monkeys, thirsty foreign trees, and—if one has a taste for the supernatural—a wide range of ghosts, spirits and demons. These migrants have been an integral part of an increasingly complex set of systems—geological, ecological, political, economic and religious (the subject matter of Chapters 1 through 5, respectively)—that have intertwined to create the Delhis of the past and the present.

The Ridge has played a crucial role in all these systems. Though its ecological functions may be foregrounded today, these are just one part of a much larger whole. The Ridge's trees can't be separated from the stones below them, nor the cities that rose and fell around them. Environmental and social history blur. Only with this perspective does a clear picture of the Ridge, and of Delhi as a whole, emerge.

A Tour through the Ridge's History

Those who focus exclusively on the Ridge's ecology usually do so to argue for the protection of its Reserved Forests. In this context, the Ridge's "pristine glory" is frequently evoked, though often more rhetorically than literally. Delhi's environmentalists recognize that, in a real historical sense, the Ridge is far from untouched. They seem to realize the shakiness of the "pristine glory" logic.

And yet this logic keeps on reappearing in discussions about the Ridge, not least in crucial court cases and powerful planning documents. The persistence of this logic suggests the necessity of refuting it, especially since the language of untouched glory is easily appropriated by the state to justify, for instance, the demolition of informal, low-income housing that has "sullied" the pristine Ridge (an all-too-common phenomenon).

A whirlwind tour through Delhi's history is enough to show the flawed logic of "pristine glory" in the Ridge; it also has the added benefit of showing the bewildering complexities of the Ridge and the city. Such a tour brings home a profound truth: there is no static baseline to which we can return. Only by recognizing this, and dropping the rhetoric of a pristine past, can the Ridge be sustained in a just way.

Our tour begins in the 1930s, when Mirza Farhatullah Beg, an author from noble Mughal stock, bemoaned the degraded state of

Delhi's environment. Beg described Mehrauli's idyllic *jharna* with great nostalgia, contrasting its glory days with its twentieth-century state of disrepair. His lament is a familiar one:

Now, the pleasantness no longer exists.... No longer does the water trickle down. The canals have dried up. The cisterns are filled with the rubble of the ruined buildings. Trees bear no fruit and most of them have been cut down. The slippery stone has broken into pieces. Only a few buildings still stand. In a few days, however, even those will be gone.[5]

Perhaps Delhi's environment was better in the nineteenth century? A report by a prominent environmental organization has traced the Ridge's fall from grace to the establishment of the British New Delhi in 1911: "with the transfer of capital of India from Calcutta to Delhi during British times, developmental activity started in the city. The Ridge... started losing [its] natural state... The perfect balance and harmony of nature had been disturbed."[6]

But it is hard to discern any balance and harmony in nineteenth-century Delhi. Most obvious are the traumatic events of 1857; in that tumultuous year, the British government, after repressing the great anti-colonial uprising with considerable difficulty, cut down trees on the Ridge and throughout Delhi. This was both a form of revenge and a way of ensuring clear lines of sight for surveying the rebellious population.

Maybe the good years were before British rule and the violence of imperial interventions? But then, in pre-British Mughal times, the pressures of urbanization in Delhi were immense, and must have taken their toll on the environment. At the beginning of the eighteenth century, the population of the city and its suburbs touched 400,000. By the end of that century, Delhi was surrounded on all sides by eight to twelve kilometers of intensive cultivation, necessary to feed the city. When the British first surveyed the land that they had taken over from the Mughals, they were dismayed to find a largely denuded landscape.

Going further back, we find that historical records of deforestation in Delhi are almost as old as historical records of the city itself. The fearsome warrior Timur cut down trees around the northern section of the Ridge in 1398, as part of his deadly invasion of the city. The great Sultan Balban cleared the forest around Mehrauli in the 1260s. And on and on.

Then we enter the murky realm of mythology. It's not very encouraging for nature lovers. Since the time of the epic *Mahabharata*, the Delhi region has been associated with tree felling, forest fires, and unrepentant animal slaughter. Consider this passage from the epic, relaying, in alarmingly celebratory tone, the role of heroic Arjuna and Krishna in burning the Khandava forest on the outskirts of the city of Indraprastha:

> *The two tigerlike men started a vast massacre of the creatures on every side.... As the Khandava was burning, the creatures in their thousands leaped up in all ten directions, screeching their terrifying screams. Many were burning in one spot, others were scorched—they were shattered and scattered mindlessly, their eyes abursting.... When they jumped out, Arjuna cut them to pieces with his arrows and, laughing, threw them back into the blazing Fire.*[7]

It is difficult to draw straightforward connections between myth and ecological history. One does not want to follow the road paved by Hindu nationalist historians and suggest that Delhi is literally Indraprastha (more on this danger shortly). However, several scholars have argued that it is possible to read the *Mahabharata* and related texts, not for literal truths, but for suggestions about how northern India was transformed during the time the epic was composed.[8]

Specifically, these grisly scenes may be allegorical references to a phenomenon affecting broad swaths of land, including the Delhi region: the use of fire to turn forests into pastures. This strategy was used by livestock-rearing groups to expand the zones favorable to their lifestyle. In doing so, they found themselves at war with those who considered the forests their homes: the hunter-gatherers who kept no cattle and tilled no fields, but roamed the land seeking new sources of sustenance from day to day. The livestock-rearers (or, to use the technical term, pastoralists) won the fight, and set the stage for the use of Ridge-as-pasture, a role that it played for many centuries, and that it still plays today, at least in isolated areas. But this bloody victory, though celebrated in the pages of the *Mahabharata*, brings us no closer to Delhi's golden past.

Perhaps it existed before the incursion of pastoralists? After all, the domestication of animals, along with the domestication of grains, ushered in the so-called Neolithic Revolution, which some see as an ecological watershed—one to be mourned, not celebrated. The scholar Jared Diamond has dubbed the Neolithic Revolution "the worst mistake in the history of the human race".[9] Besides the ecological consequences of the revolution, Diamond points to other undesirable consequences: plummeting quality-of-life indicators for human communities; the slow development of social classes, and oppressive states to rule those classes, all made possible by the unequal hoarding of agricultural surplus; and an increasingly uneven division of labor between the sexes.

Other scholars, though, have questioned Diamond's fall-from-grace narrative of the Neolithic Revolution. Some doubt whether it was really even a revolution at all, since the ill effects that Diamond catalogs do not necessarily follow from the domestication of plants and animals, at least in any immediate sense. For instance, with regards to economic inequality, "while agriculture allowed for the *possibility* of more unequal concentrations of wealth, in most cases this only began to happen millennia after its inception."[10] Humanity's ancient history, ecological and otherwise, was marked by contingency, variability and a diversity of paths, with no easy generalizations made about uniform decline. Taking the case of gender in the ancient Middle East, for example, David Graeber and David Wengrow argue that some "Neolithic societies look strikingly egalitarian when compared to their hunter-gatherer neighbours, with a dramatic increase in the economic and social importance of women, clearly reflected in their art and ritual life."[11]

What, then, was the case for Delhi's hunter-gatherers, the ones driven away by pastoralists and their ilk? We have little evidence to go on, besides the Paleolithic stone tools that will be discussed in the next chapter—but the sheer profusion of these tools suggests a complexity and intensity of production that belies the Edenic idea of a people in easy harmony with nature. Further, we should not assume that the pastoralists disrupting life in northern India were the only ones who harnessed fire to transform the environment; on the contrary, there is evidence that hunter-gatherers have used fire to shape the landscape for at least 400,000 years.[12]

Finally, then, we could look to the pre-human past to identify Delhi's golden days. Certainly, in the present age of environmental crisis, there is an urge to idealize earthly life sans humans; this, in fact, is the premise of the wildly popular book *The World Without Us*.[13] And a long lineage of environmental thought has placed humanity outside nature, celebrating the harmony of natural systems and ruing their disruption by ill-considered human intervention.

But here too we run into problems. Even before humans spread around the globe, nature did not exist as an undisturbed whole, in perfect balance. To take just one apposite example, the formation of the Aravalli mountain range 1.5 billion years ago was hardly a balanced, peaceful process. It was a dynamic, radical change in the landscape, full of violent ruptures and volcanic flows and unexpected metamorphoses.

The Ridge has long been the site, not of pristine glory, but of "discordant harmonies", to borrow a term introduced by ecologist Daniel Botkin.[14] A belief in "the balance of nature" seems deeply ingrained in the human psyche, but Botkin insists that this metaphor is increasingly incongruous with the findings of ecology, and needs to be set aside. Instead, we need to accept that many species, including humans, actually thrive on change, and that a complex, unbalanced, sometimes random set of processes has nonetheless produced remarkable harmonies and stunning diversities.

Following Botkin's advice, we can abandon the search for Delhi's idyllic past. But recognizing the ever-changing nature of Delhi's environment does not mean accepting uncritically all human interventions in Delhi's landscape. Asserting the inevitability of change does not mean blithely assenting to *all* change. As the evidence of deforestation in Delhi suggests, humans have, for many centuries, introduced changes that have devastated the region's ecosystems—changes that were far too fast and far too reckless, and that thus cut off the possibilities of creative adaptation that have characterized more positive environmental change.

Neolithic Revolution   Neolithic Revolution   Stone Age   Hadean Eon
Agriculture   Pastoralism

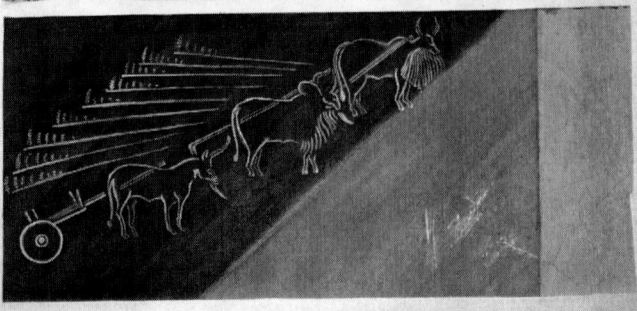

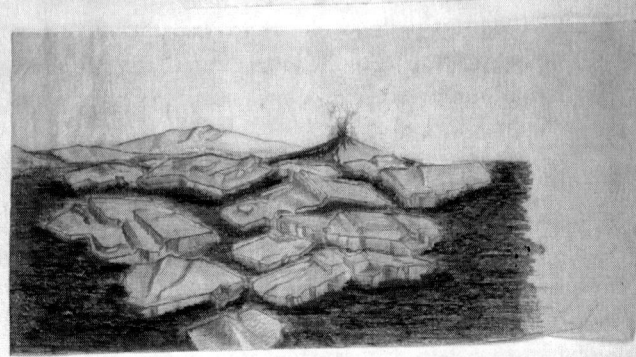

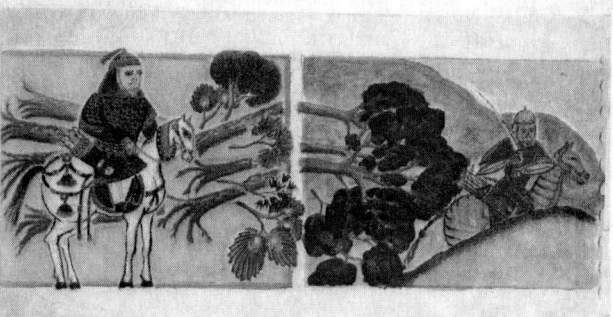
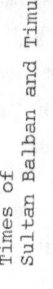

Times of
Sultan Balban and Timur

Mughal Period

British
Colonial Period

The Present Day

No Pristine Glory

But not all changes in Delhi have been so destructive. In the fourteenth century, the sultan Firoz Shah Tughlaq planted trees around his hunting lodge on the Northern Ridge; centuries later, the Mughals created elaborate, lush gardens and networks of streams and wells to revive a parched city; and both the British and post-colonial governments embarked on massive reforestation efforts. The very existence of the modern-day Ridge forest, after so many centuries of deforestation, points to the malleability and resilience of the landscape. It also suggest that humans are, for better or worse, an integral part of an ever-changing, dynamic set of ecologies.

Pristine Glory of Another Kind

Perhaps I am so wary of the "pristine glory" logic because it has uncomfortable, if largely unintentional, resonances with another kind of nostalgia: not environmental, but religious. These two brands of nostalgia employ strikingly similar metaphors. In both versions, there is, somewhere in the distant past, a unified, balanced whole, and this utopia is broken into pieces by an unwanted intruder.

In the ecological version, this intruder is the human race, or the agriculturalist, or the imperialist, or the capitalist (depending on who is telling the story). In the religious version, the intruder is the Muslim.[15] This, at least, is the history advanced by strident Hindu nationalists, who nurture the idea of a harmonious past when Delhi was free from turmoil and conflict. In this view, Delhi is equated with mythical Indraprastha, and the downfall of Delhi is pegged squarely on the year 1192, when the Hindu King Prithviraj Chauhan was defeated by the Muslim invader Mohammad Ghori. This is one of the ignominious defeats that Prime Minister Narendra Modi is alluding to when he intones—as he often does in his speeches—that India has endured more than 1,000 years of slavery.[16]

At present, the Hindu nationalist view of Delhi is not the dominant one; Delhi is a popular tourist city, and its most visited sites are the intricate architectural marvels from the days of Muslim rule. And yet, in these days of ascendant Hindu nationalism, there are worrying signs. In December 2014, soon after the Hindu nationalist Bharatiya Janata Party (BJP) came to power, M. Venkaiah Naidu, the

Urban Development Minister, suggested that Delhi's official name be changed to Indraprastha. Such recent moves have historical precedent, sometimes overlapping in uncomfortable ways with the history of the Ridge and its preservation. The most striking example of this is Jagmohan, the iron-fisted bureaucrat who expanded the Ridge areas under state protection in 1980 and who, later in his career, promoted an increasingly aggressive glorification of Delhi's Hindu history, including the construction of a statue of Prithviraj Chauhan in one Ridge park in the early 2000s.[17] (More on him in Chapter 3.) In this pursuit, he was aided by the then Home Minister and powerful Hindutva leader, L. K. Advani, who inaugurated the Prithviraj Chauhan statue, and later launched a spurious attempt to claim the Qutb Minar as a Hindu monument.

A nuanced history of the Ridge undermines this Manichean understanding of Delhi and its pasts, as we will see. Just as Delhi's ecology has been complex, fragmented, and ever-evolving, so too have its human societies, from their religious practices to their social norms to their economic strategies. Hinduism has not been a timeless, stable entity, nor has Islam, nor any of the other faiths that have mixed in Delhi's soil—sometimes in violent conflict, sometimes in peaceful syncretism. Exploring the social and ecological processes that have unfolded on the Ridge can help foreground Delhi's religious diversity.

Anti-Ecological Environmentalism

If one of this book's main aims is to help the Ridge regain its multifaceted history, the other is to situate it more firmly in its geography—or, more precisely, to sketch out its interconnections with local, regional, national and global geographies. Both these aims are animated by the same purpose: to broaden the way we think of Delhi's environment, and to reconnect the humans of Delhi with the ecological lifeblood of their city.

Just as a narrow insistence on the Ridge's "pristine glory" dilutes the region's complex history, a narrow focus on "clean and green Delhi", a favorite rallying cry for the city in recent years, diminishes the Ridge's complex dependencies with human and non-human systems that span the city, the country and the world. This myopic view obscures the root causes of Delhi's environmental crisis, and, even worse, ends up blaming those who least deserve it.

My meanderings through dystopian Mehrauli could, at first glance, support this view: just next to a dilapidated basti was a garbage-filled, sewage-choked stream. Were the basti's inhabitants responsible for the pollution? This, in fact, is the knee-jerk response of much of Delhi's elite, deeply shot through with class and caste bias: the poor are the dirty ones. At best, this leads to the conclusion that the poor need to be uplifted, both for their own sakes and for the sake of the environment (this is usually what the government says it will do). At worst, it leads to the conclusion that the poor are irremediably dirty and polluting, and must be removed from the city by any means necessary (this is usually what the government actually does).

This has been the dominant way of thinking about Delhi's environment, especially in the post-Independence period. Over the past several decades, there have been many waves of demolition drives, which have swept away hundreds and thousands of poor Delhi residents. At least since the 1990s, these demolitions have been justified with explicitly environmental rhetoric. To give just one prominent example: in April 2004, more than 150,000 people were uprooted, their homes destroyed, because they lived on the banks of the Yamuna River, and, according to the Delhi High Court, were responsible for dumping sewage water into the river.[18]

But who is really responsible for polluting the Yamuna? A study by the non-profit organization Hazards Centre showed that the demolished settlement had only been responsible for a scant 0.5 percent of the effluent discharged into the river; the vast majority of pollution came from 22 drains that dumped untreated sewage into the Yamuna.[19] These drains snake through much of Delhi, collecting waste largely from middle-class and upper-class residential areas. If the poor are dirty, it's because they're forced to live in the filth that the rich have created.

As with the Yamuna, so too with the Ridge. The city's "green lung" is often touted as a carbon sink, which can mitigate the ill effects of air pollution and the production of climate-altering emissions. Following this logic, many settlements on the Ridge, deemed encroachments, have been bulldozed into oblivion. But what is the root cause of Delhi's air pollution? Road transport is responsible for more carbon emissions than any other sector. But vehicular pollution is the responsibility of a relatively small, elite section of the city's population; the 2,000,000 cars on the road belong to only 20 percent of the city's households.[20]

The hypocrisy of a shallow environmentalism, which demolishes slums while promoting car ownership, has been noted by many. Academics analyzing this phenomenon have given it many names: bourgeois environmentalism,[21] anti-poor environmentalism,[22] middle-class environmentalism,[23] and so on. Conceptually, its chief flaw is that it severs the connection between one's own consumption and one's immediate environment, so that it becomes perfectly reasonable for a car owner to demand the demolition of poor settlements on the Ridge. This thinking is, at its core, deeply anti-ecological, since it ignores one of ecology's basic tenets: that everything is interconnected.

When this profound truth is ignored, the Ridge becomes an isolated space, cut off from the rest of the world, to be protected by further strengthening the barricades around it. Sometimes, this is quite literal: there has recently been a push to build walls around several parts of the Ridge. At other times, this logic becomes farcically superficial; the Delhi government has resorted to erecting bamboo screens along major roads during international events in order to hide "unsightly" low-income settlements and create a pleasing "natural" view of the city for visitors, cut off from the reality of those who toil to make the city run.

The Economics of Disconnection

This way of thinking is not just due to greed, or laziness, or narrow-mindedness. The obscuring of connections (ecological and otherwise) is a structural feature of today's international economy, and can argu-ably be traced back much further, to the development of social classes many thousands of years ago. The elites in early class-based societies no longer had to labor to produce their everyday means of subsistence, and thus lost a vital, visceral connection with nature. It was these elites who ruled, and wrote, and presided over religious rituals, thus shaping a culture that was increasingly distanced from nature.

In a capitalist economy, this sense of disconnection is amplified considerably, since even those who work to produce the necessities of life are generally not toiling to create their *own* means of subsistence. Rather, they are producing commodities that will be sold by someone else, to someone else, on the open market; at the same time, they are using their (often meager) earnings to buy the goods they need from the market. This process automatically creates a disconnect between the commodity's producer and its eventual consumer.[24]

In contemporary times, this disconnect has become truly extraordinary, as high-tech products are assembled in factories scattered across the globe, with supply chains stretching across several continents. It's easy to miss the dark side of these networks. When we use our smartphones, we may, in some abstract way, know that our technology is the product of sweatshop labor in China, environmentally-scarring mining in South Africa, and so on, but it's very difficult to feel these connections viscerally; the system itself has distanced us from them.

Our sense of disconnection from nature is, in one way, an illusion. But this illusion is propped up by economic, political and social realities, which have their roots in the birth of class societies and which have intensified immensely under global capitalism.[25] For much of human history, this has led to an arrogant triumphalism: humans have defeated nature or have escaped the bonds of nature. (Of course, someone must be working with nature to produce food and build homes, but these people are pushed out of the popular imagination, with serious consequences: is it any coincidence that many victims of displacement in Delhi and beyond have been farmers and construction workers?)

Today, in the age of global environmental crisis, one hears this triumphant language less. Ironically, the very technology that was once used to tout humanity's separation from nature is increasingly being seen as evidence that humans, in the end, can't escape nature. If we keep adding carbon dioxide to the atmosphere, we will be bound to a world of rising temperatures, no matter how "free" from nature we may temporarily feel. Nature will have the last laugh.

Despite the increasingly apparent inseparability of humanity and nature, some still fall into the trap of seeing nature as a totally

The Alienation of Labor

separate realm, which must be cordoned off from humanity's poisonous influences. Though such advocates are resolutely opposed to an arrogant ideology of conquering nature, they duplicate the binary thinking at the heart of that ideology.

This happens all too often with the Ridge. Small areas of the Ridge are cordoned off, and then Ridge defenders are incensed when the cordons are overrun. The frustration is understandable, but it doesn't seem to take into account that Delhi is a city with a severe housing shortage (especially for low-income populations), a commitment to driving up land prices (encouraged by both the public and private sector), and a never-ending inflow of migration (supported by an explicit government policy of promoting urbanization).

This is not to say that Delhi doesn't need green spaces, nor that some wild zones shouldn't be left relatively untouched by humans. But merely creating islands of conservation whilst doing little to address the interconnected economic and political causes of ecological destruction, will inevitably lead to these islands being washed away, sooner or later. In the case of the Delhi Ridge, in the middle of a frenetic megacity, it seems likely that wave of destruction will come sooner rather than later.

Darkness and Light

I did not come into the Ridge project with such pessimism. The seeds for this project were planted far more playfully. I moved to Delhi in 2010, and soon after moving into a flat near Delhi University, I began hearing alarming rumors about a nearby park.

Muggings. Murders. Mystics. Monkeys.

Curious, I began wandering around Kamla Nehru Park on the Northern Ridge, and to my eyes, it seemed to be quite tame (except for the ubiquitous monkeys). It was a public park with big wide paths, some of them even paved. Its neatly manicured hedges kept the more unkempt thickets of trees firmly in the background. It was filled with diligent joggers and disciplined power-walkers. But the intense public perception of the Ridge kept drawing me back. There must be something, I figured, that gave the Ridge such an outsized place in the imagination of those who lived near it, many of whom would not dare to venture into it.

The Ridge, I slowly discovered, was, for many, a place to seek transcendence, sometimes a profoundly spiritual one, sometimes a more mundanely material one. It was a place to escape the constraints and conventions of a restrictive society. Furtive lovers met there, paying off security guards and disappearing into the bushes. Rebellious high school students went there to smoke pot and drink beer. Sadhus and yogis also smoked pot there, in search of a more divine high (or so they said).

In a strange way, the present-day "uncivilized" wildness of the Ridge reflects much older associations of nature and transcendence— even if this wildness is just a perception, ignoring all the careful horticultural work that has contributed to the Ridge's current incarnation. This way of viewing the Ridge is a double-edged sword. It reflects the continuing divide of human culture and wild nature. The Ridge parkland is only seen as a place of transcendence because it is so starkly opposed to the everyday life of conventional society.

And yet, in the flickers of sacredness, there is some sense that this is our ultimate home, that the Ridge is expressing some truth that our society has forgotten. This is, perhaps, a grandiose way to describe a park full of lovers, joggers and stoners. But, as we shall see, the Ridge has been home to miracles and tragedies, gods and demons, deaths and resurrections, for many centuries. They point to a deeper reality.

It is a reality, though, that cannot be disentangled from the Ridge's economic history, nor its political history, nor its environmental history. The spiritual life of the Ridge does not exist in a vacuum. The same can be said of Delhi, whose existence as a city depends crucially on the stones and soil of the Ridge. The story of human society and the Delhi Ridge is, to a large extent, the story of the growing rift between the two, and the way that rift has expressed itself through statecraft, surplus accumulation and spiritual belief. But, reading history against the grain, the story also provides clues about how this rift can be healed.

1 STONES
Shifting Geologies of the Ridge

A Walk through (Pre)History

In 1986, an archaeologist named S. S. Saar made a startling discovery. Walking in the middle-class South Delhi neighborhood of Malviya Nagar, he noticed a mysterious glint in a pile of sand. Mixed in with the sand, which had been dumped there by a crew of construction workers, he found a handful of ancient stone tools, carved from quartzite.

This inspired another archaeologist, A. K. Sharma, to find the source of these tools, and his search eventually led him to Anangpur, a village just beyond the southern boundary of Delhi, where large-scale quarrying operations had been initiated as part of larger efforts to mine the Ridge. A series of archaeological expeditions followed, and the research teams found thousands of stone tools scattered through the area, evidence of a large Stone Age site of habitation.[1] Despite the fact that many stone tools had been destroyed or misplaced in the process of mining, Anangpur remains one of the largest Stone Age sites discovered in India.

If not for the construction boom in Delhi, and the need to dig up the Ridge's rocks and sands for building, Delhi would still be ignorant about the largest settlement of its earliest human inhabitants. Those intent on building a new Delhi couldn't help but dig up the past, quite literally. And all that digging inevitably circles back to the Ridge, the source of the city's geological riches.

Geology has played a pivotal but often overlooked role in the development of both historical and present-day societies, and Delhi is no exception. Geology underlies our lives, not just physically, but economically, socially, and technologically. Consider, for instance, the absolute centrality of fossil fuels to modern life. The industrial revolution was propelled by steam power, which depended on the fossilized remains of vast coal deposits from the Carboniferous (literally, "coal-bearing") Age, about 300 million years ago. Similarly, the political economy of the twentieth century hinged on the geological remnants of the Cretaceous Period 100 million years ago, when vast gas and oil reserves formed due to a lack of oxygen in the deep sea. The geology of Delhi may not have such global resonance; nonetheless, every society in Delhi's history has drawn on the unique geology of the region, from the Stone Age to the present day.

Those who ignore Delhi's geology do so at their own peril. In September 2014, newspapers reported yet more delays in the construction of the city's "Signature Bridge", which was meant to be completed for the 2010 Commonwealth Games.[2] The culprit this time: an "unexpected rock profile" on the riverbed. Bridge, meet Ridge. Anyone familiar with Delhi's basic geology would know that the above-ground outcrops of the ancient Aravallis peter out precisely at the section of the Yamuna where the bridge was being built. But the Aravallis keep on going underground, all the way to Haridwar, covered by relatively recent alluvial deposits from the river systems of the "young" Himalayas. The rock profile found by the bridge engineers should not have been so unexpected after all.

Though the stones of the Ridge form the foundation of much of Delhi, they remain largely hidden from view. They are buried underground (or underwater) or hidden by shrubs and thorns, with only the occasional boulder visible on higher ground. What follows in this chapter is a work of excavation, unearthing the ridge-ness of the Ridge. For these stones have a story to tell, not just about geology, but about the hunter-gatherer prehistory of Delhi, the bloody travails of the medieval period, the rise of imperial bureaucracy, the struggles of the city's workers, and the dreamy imagination of the city's elite.

I am hardly the first to be interested in these stones. Delhi has long been home to people whose lives and livelihoods depended on the region's geology, from our hominid ancestors who first found tool-worthy material in eroding stones to the modern-day contractors selecting the best type of sand for making concrete. They are all geologists in the broadest sense of the term; their understanding of the Ridge's rocks has pushed forward scientific and technological advances and pioneered new methods of using Delhi's natural resources, in both constructive and destructive ways.

The Stones Come Alive

For the Paleolithic inhabitants of Delhi, one stone would have held particular importance: quartzite. This is the stone from which so many tools were crafted, the stone that S. S. Saar stumbled upon so fortuitously. Quartzite can be found in abundance throughout the Aravalli range, which begins in Gujarat, traverses all of Rajasthan, and ends with the low hills of the Delhi Ridge. Once a grand mountain chain, the Aravallis have been subject to erosion and gradual weathering over the course of more than a billion years. However, quartzite is particularly resistant to erosion, and so it remains prominent in many parts of the range, including the Ridge. It would have been a beacon to early tribes.

Although these tribes seem to exist in the farthest reaches of the ancient past, they are of remarkably recent vintage when compared to the quartzite they were so adept at shaping. Quartzite appeared to the tribes as a fixed, permanent substance, with a solidity and heft that proved useful for many purposes. Indeed, this is how quartzite appears to us today. But when considered on a sufficiently vast timescale, the stones of the Ridge (quartzite and all the others) take on a life of their own, clashing and clanging and eroding and re-forming over many millennia.

Before zooming in on a more intimate, human-centered timescale, it is worthwhile to meditate for a moment on the unfathomable expanses of geological time. The quartzite of the Ridge is 1.5 billion years old, the same age as the Aravalli Range itself. It is 7,500 times older than humanity and 150,000 times older than the birth of agriculture. No wonder it seems so constant to humans.

To understand how quartzite formed, and why it has properties that have been so useful to Stone Age and modern humans alike, we'll have to take a brief plunge into the forbidding, Greek-inflected jargon of geologists. For instance, the oldest time period on the geological time scale is called the Hadean eon, named after Hades, Greek god of the underworld. The name is meant to evoke the fiery, "hellish" conditions on Earth at the time, as much of its surface was still molten; the technical term is "extreme volcanism". In those times, there was more lava than rock, and the Ridge was still waiting to be born.

The oldest rocks in the Aravallis are around 3.3 billion years old, significantly predating the formation of the mountain chain. This marks them as products of the Archean eon, when the Earth was beginning to cool. Appropriately, the term "Archean" comes from the Greek word for "beginning" or "origin"; Earth's geology, including our little corner of it, the Ridge, properly starts in this eon. After the end of Hades' reign, various continents began to take shape.[3] The basic constituents of these continents were cratons: stable, thick pieces of the Earth's crust.

From around 3 to 2.5 billion years ago, there was a period of rapid thickening in the craton that now hosts the Delhi Ridge. Evidence of this can be found in an analysis of the present-day Aravalli mountain range, which reveals an ancient formation at its base, dominated by granitic rocks. During the Archean eon, the crust in the Aravalli region was at least twenty kilometers thick.

The next eon, the Proterozoic, was when the Ridge as we know it took shape. Geology is not a field known for its ironies, but it is surprising to learn that the mountains of the Ridge actually began as a basin.[4] Although the immediate cause of the Aravallis' formation was the pushing together of land, it has its origins in the pulling apart of the Earth's crust.

This process, known as rifting, began approximately 2.5 billion years ago and continued until about 1.9 billion years ago. The craton that had grown so thick in the Archean age of beginnings began to thin out in the Aravalli region, as tectonic forces started pulling it apart. Fiery heat returned to the Ridge, as magma burst through the cracks in the earth. Due to all this thinning out, a large basin formed, which began to collect volcanic rocks, as well as sedimentary ones like sandstone.

Rifting was followed by about 300 million years of cooling as the crust settled back into an equilibrium. But this equilibrium would not last long. The main event was about to start: the "Delhi orogeny". On a normal human timescale, this event would hardly be perceptible. A human would not find much change in the Aravallis even during its most drastic periods of transition. But if we speed up our timeline, and take in a billion years at a time, we can see how transformative the Delhi orogeny was.

"Orogeny" refers to the collision of huge masses of rock and the forming of mountain chains. When textbooks say that the Aravallis are 1.5 billion years old, this is what they mean: the Delhi orogeny took place about 1.5 billion years ago. Two pieces of earth that were formerly being pulled apart to create basins were now being pushed together. All the sediment got pushed up as the mountain range was formed. In describing the dramatic process of rift inversion, the staid, technical language of academic articles slips up and betrays some emotion; in orogeny, part of the basement rock "suffers" from deformation and metamorphism.

That last hardship (metamorphism) is particularly important for understanding the current geological make-up of the Ridge. As the basins were pushed together and began their ascent, the volcanic and sedimentary rocks that were previously present in the region changed into metamorphic rocks because of the heat and the pressure of the mountain-making process. And so, the sedimentary rock known as sandstone was heated, compressed, and transformed into quartzite.

After this massive event 1.5 billion years ago, the Aravalli range continued to evolve, as any geological phenomenon does; smaller basins emerged along the flanks of the mountains, and new sequences of sedimentary rock formed on these flanks. A few dramatic geological surprises were yet to come, as we'll see. But the basic form of the Aravalli mountain range had been established, and the key geological components of the Ridge were in place, including the quintessential metamorphic rock of the area: quartzite.

Enter Hominids

Over the course of the Ridge's history, humans have engaged in a slow, and then alarmingly rapid, process of resource extraction, chipping away at stones that had come into being over the course of billions of years. In one sense, humans only accelerated processes that were already well underway; the Aravallis had begun eroding well before humans entered the scene. But again, we must weigh geological time-scales against more anthropocentric ones. The pre-human erosion occurred at the glacial pace of geological time, the forces of wind and of water weathering away the rock molecule by molecule. It was a slow

Hadean Eon *4 billion years ago*

the earth is
in a volatile
state.

Archean Eon *3.3 billion years ago*

the earth cools

continents
form

the rock of the
Aravalli forms.

Proterozoic Eon *2.6 to 1.9 billion years ago*

rifting occurs in the Aravalli bedrock.

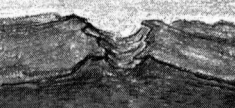

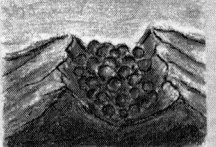

The Delhi Orogeny *1.5 billion years ago*

pushed together

under
intense
pressure

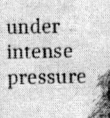

sedimentary
rock becomes

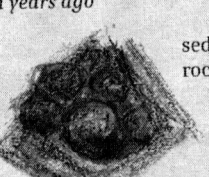

Quartzite.

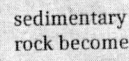

1 billion years ago

intense pressure,
over several years
creates Quartz.

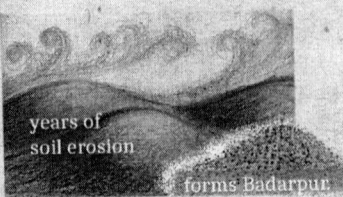

years of
soil erosion

forms Badarpur

Formation of the Ridge

Homo Habilis *2.5 to 1.3 million years ago*

 in the Rift Valley  hills afford shelter and quartzite for making  pre acheulian tools.

Homo Erectus *3.5 to 1.2 lakh years ago*

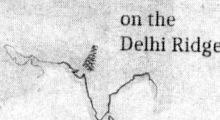 on the Delhi Ridge  quartzite is made into  acheulian tools.

Homo Sapiens *1 lakh years ago*

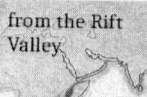 from the Rift Valley along coastlines,  and rivers, homo sapiens reach the Ridge.

Homo Sapiens *60 to 70 thousand years ago*

 using quartz they make 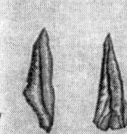 microliths creating effective weaponry.

quartzite and badarpur form many Delhis.

Hominid Migrations

disintegration imperceptible to human eyes. Human intervention happened much more quickly, and thus its impact has been much more dramatic.

The first signs of this intervention are remarkably ancient, though their discovery has been remarkably recent. The massive archaeological site at Anangpur, for instance, was unearthed only in the late 1980s. These recent findings have provided a glimpse into the Stone Age pre-history of Delhi. The very term "Stone Age" points to the centrality of rocks to the everyday life of ancient communities. And fittingly, the only evidence we have from that period in Delhi is a growing set of stone tools dug up by archaeologists (or construction crews).

It is surprising how little these Stone Age remnants figure in popular histories of Delhi. Part of this may be due to the notorious difficulty of interpreting such ancient, scattered evidence. Part of this, perhaps, is political. Hindu nationalist narratives have given Delhi thoroughly Indian (by which they mean thoroughly Hindu) roots. Such narratives are quick to label Indraprastha, from the epic *Mahabharata*, as the first significant habitation in the Delhi region, despite the questionable historicity of the epic.

Indraprastha must be the first, the origin, the precursor and progenitor of all other Delhis; the "sons of the soil" Hindutva narrative demands this. No matter that the *Mahabharata* itself implies that the area was already inhabited before the heroic Pandavas made their entrance. And, more relevant for our purposes, no matter that there is tangible, hard-as-rock evidence that, many thousands of years before the already-misty past of the great Bharata war, other migrants had already made their way to the Delhi region, with an origin story more universal in nature.

This evidence comes from at least forty-three sites around Delhi, which are detailed in a landmark 1987 report on the city's Stone Age past.[5] These sites are, unsurprisingly, clustered around the Ridge (where else to get stone tools?) and span a vast period of time, from the lower paleolithic (more than 200,000 years ago) to the meso-lithic (as recently as 5,000 years ago). The 1987 report built on earlier, more fortuitous discoveries, including the very first discovery of ancient stone tools in 1956, when a scholar happened to find four such tools near the main gate of Delhi University in the north of the city. Several other early finds took place in the campus of Jawaharlal Nehru University (JNU) along the southern edge of the Ridge. More than one archaeologist has commented that it is embarrassingly easy to chance upon Stone Age relics when strolling in Delhi's hills.

Many of the ancient stone tools discovered in the Ridge, including those of Anangpur, are categorized as "Acheulian", a name that comes from the town in France where such tools were first discovered. Such tools were widely used in several continents for hundreds of thousands of years, with the oldest traced to 1.7 million years ago.

When we venture this far back in history, the meaning of the word "human" starts to get a bit shaky. Were the authors of the massive stone industry in Anangpur and throughout the Ridge really human? It is unlikely that they were *Homo sapiens* or, to use a favored archae-ological term, "anatomically modern humans". But they belonged to the genus *Homo*, Latin for "human", and in the archaeological liter-ature, all Stone Age sites are broadly considered part of early human life. (For instance, there's the common claim that "Acheulian tools were the dominant technology for the vast majority of human history.") So how human, exactly, were the Ridge's earliest hominid inhabitants?

The simple answer is that the Ridge's Acheulian tool-makers belonged to the species *Homo erectus*, a direct ancestor of *Homo sapiens*, and similar to our species in several crucial ways. They had significantly larger brains than the very first of our genus, *Homo habilis*, and shed some bulk in order to move more nimbly. Also, as compared to *Homo habilis*, the physical differences between males and females were less marked. If not modern humans, they were well on their way to being so.

But things are rarely simple in the study of the prehistoric past. Some archaeologists suggest that *Homo erectus* is actually an archaic subcategory of *Homo sapiens*. Others insist that *Homo erectus* is not the direct ancestor of *Homo sapiens*; that honor goes to *Homo ergaster*, a species whose very existence is hotly debated. Adding to the confusion is the difficulty of connecting the general debate on human origins with the specific—and scant—evidence from the Delhi region. The only clues in the area are stone tools; no fossilized skeletons have yet been found, making it all the more difficult to determine just what kind of hominid created the tools. And the dating of these tools is highly imprecise. A recent study of Acheulian tools found on the JNU campus could only give the extremely rough estimate that the Stone Age site was somewhere between 128,000 and 350,000 years old.[6]

Yet even this estimate is helpful. There is a general consensus that *Homo sapiens* only reached India about 70,000 years ago, well after the Ridge had its first hominid inhabitants.[7] This makes the *Homo erectus* label seem appropriate for the earliest Ridge settlers, as does the fact that, in other parts of the world, Acheulian tools are often found with the skeletal remains of *Homo erectus*.

The species *Homo erectus* originated in Africa, the birthplace of our genus. Given our geological concerns, it is worth noting that *Homo* did not originate just anywhere in Africa. Our genus was born in the Great Rift Valley, a set of broken hills and surrounding savannas that stretched from the Red Sea to present-day South Africa. This immense ecosystem was well-suited for early hominids largely because of its hills and caves, which served many purposes: as a place to hide from dangerous predators, as shelter from the elements, and a vantage point for spotting prey.

The earliest of our genus, *Homo habilis*, with its pre-Acheulian tools, likely relied much more on scavenging than on hunting, and the species stuck to Rift Valley-like environments even when some groups migrated out of Africa. But by the time *Homo erectus* had emerged, our genus had shed some of its original constraints. The harnessing of fire and the use of Acheulian tools allowed these hominid societies to roam more freely over more varied terrain, and to hold their own against a wide set of predators. Still, even for *Homo erectus*, hilly areas would have been a clear draw, both because of the stone they provided and the vantage points they afforded.

This helps explain why there are such dense clusters of Stone Age remains in the Delhi Ridge. The broken, meandering hills of the Ridge would have provided an ideal environment for hunting, and not just because of the ease of spotting prey. The archaeologist A. K. Sharma speculates about the hunting methods of Delhi's earliest inhabitants and their use of escarpments or steep slopes: "Escarpments... provided ideal spots for the paleolithic man to drive herds of less harmful animals towards escarpments and force a fall to kill or injure them grievously."[8]

For these early hunter-gatherer communities, the tools fashioned from Ridge rock were essential for tasks like hunting, starting fires, cutting up the bodies of dead animals, digging up roots, and piercing the shells of hard fruits. The use of Acheulian technology also suggests considerable intelligence and knowledge passed on from generation to generation. Paleolithic inhabitants of the Ridge would have recognized the variations of quartzite in the hills, and would have selected those blocks of quartzite that were most erosion-resistant, and thus best suited to tool-making.

The Acheulian tool-makers were encountering a rockscape that was highly weathered. This is not only because of the age of the Aravallis, but also because the stone in this region has unusually high amounts of iron impurities, which accelerate the weathering process. Over the course of millions of years, drops of rainwater have worked their way into microscopic cracks in this iron-rich rock, slowly but relentlessly prying open the cracks and wearing away the stone. Ridge rock formations thus typically have a gray, unweathered core, surrounded by weathered white rock, with an outer covering of severely eroded, crumbling red rock. Stone age people sought out the strong gray core for tool-making.

The site at Anangpur, which has thousands of finished artifacts as well as tools in various stages of production, suggests that the settlements here were quite large. Many archaeologists have argued that the quartzite tools, aside from their various utilitarian applications, had high social value and were used as markers of social status. Some of the tools seem unnecessarily elaborate, which suggests a larger symbolic function. On this theme, speculation is rife. Archaeologists Marek Kohn and Steven Mithen have suggested that elaborate stone tools were used to impress and woo potential mates.[9] Another archaeologist, Mimi Lam, has asserted that Acheulian tools, durable and produced in huge numbers, served as the first commodities.[10] Love and money: it seems *Homo erectus* hunter-gatherers weren't so different from us after all.

The Long Journey of the Ridge's First "Modern" Migrants

While all this may be speculation, it strongly suggests that even for the earliest hominid settlers of Delhi, the stones of the Ridge were loaded with symbolic import. If this was true for the *Homo erectus* inhabitants, with their Acheulian tools, it was even more pronounced for the next round of settlers: *Homo sapiens*, our own species reaching Delhi at last.

They, or their ancestors at least, had come from rather far away. Like *Homo erectus*, modern humans have their origins in Africa. Our species first began migrating out of Africa about 100,000 years ago, and reached the Indian subcontinent around 60,000 years ago. Archaeologists speculate that India was an important early stepping-stone in the dispersal of *Homo sapiens* populations throughout Asia and Europe.[11]

These early human communities likely hugged the coast as they made their way from Africa to India and then onwards as far as Australia, traveling both by land and by sea. A recent scientific article paints an evocative portrait of this process. Migrating groups would have stuck to coastal habitats and their associated range of exceptionally rich, diverse, and economically reliable marine food resources (fish, shellfish, crustaceans, sea mammals, sea birds, etc.). They thus minimized the need for new economic and technological adaptations as they moved from one coastal location to another—in effect a process of repeated "beach hopping" from west to east, driven by steadily increasing population numbers.[12]

Things got difficult when some of the groups decided to explore the interiors of the subcontinent, likely following the rivers that flowed into the sea from the coastline. Faced with a different environment, a different set of predators and prey, a different climate and a different landscape, these early explorers had to improvise and develop a new set of survival skills and social structures suited to the new challenges they faced. This process was a slow, halting one, taking place over a period of several thousand years, but it led to a proliferation of human settlements throughout the subcontinent, including, eventually, the Delhi region. In the process, the communities relied on a crucial technology that had originally been developed in Africa: microliths, or small stone tools that were far more advanced and versatile than the Acheulian tools of the *Homo erectus*, thus furthering our genus's transition from scavenger to hunter.

Many microliths have been found in the Delhi area; they make up a significant portion of the Stone Age tools found in the Ridge. Whereas the *Homo erectus* settlers of Delhi were most concerned with quartzite, the *Homo sapiens* who first came were after another type of rock: quartz. (Yes, the two are different; yes, this is confusing.)

The quartz was a product of a surprising development in the geological history of the Ridge. Long after the Aravallis had formed and stabilized, a layer of molten magma from deep within the earth bubbled up to the surface and hardened to form rock. These formations pushed into the Ridge rock, creating veins of intrusive quartz stone that streak through the old topography. Some of these veins are clear and glassy; others more milky and turbid. All varieties were employed by the early human (that is, *Homo sapiens*) inhabitants of Delhi to make microliths.

Although microlithic tools could serve many functions, arguably the most important was their use in hunting. During this era of human history, interchangeable microliths were mass-produced, and they represented an advance over Acheulian tools in part because of their lightness and the ease with which they could be replaced on weapons. Archaeologists speculate that weapons like spears would have featured long rows of microliths, which would have pierced prey more thoroughly and caused rapid blood loss.

The increasing sophistication of microlithic tools, and with it, the increased importance and effectiveness of hunting, had a profound impact on human life in Delhi and beyond. It is difficult to trace this impact with any precision, and, once again, research on it is largely speculative. But the clues that we get are fascinating.

For one, microlithic technology may help explain why South Asia is so densely populated today. Of course, much of this region's population explosion is quite recent, but it has surprisingly ancient roots. In the period between 45,000 and 20,000 years ago, most of humanity lived on the subcontinent. This period saw three interrelated trends: population increase, environmental deterioration, and micro-lithic innovation. The environmental deterioration was not caused by humans, at least initially; rather, it was due to the advent of an ice age, which curtailed the monsoon and led to increased desertifica-tion. In response, human populations became more densely clustered in the few areas that remained suitable for habitation.[13]

Increased population density put pressure on these societies to produce more food, which in turn led to experimentation and inno-vation with stone tools. With more efficient, more lethal tools, popu-lations could not only keep their density, but start expanding again, especially when environmental conditions became more favorable. This, perhaps, is the prehistory of the South Asian population boom.

This general picture is useful for contextualizing the appearance of the first *Homo sapiens* of the region, but when we zoom in, there are very few details about the particular practices of the Ridge's hunter-gatherers. This is true about both *Homo sapiens* and their predecessors, *Homo erectus*. The scattered bits of evidence we have about Stone Age life are tantalizing; they suggest a flourishing of life in Delhi during both paleolithic and mesolithic time periods, but they give only the faintest hints of a timeline for this life, and only the vaguest sense of its shifting technologies and demographics.

Some hints are particularly suggestive. Recently, simple rock carvings were found on the Ridge, near clusters of microliths on the JNU campus. This suggests signs of a Stone Age culture in Delhi, of artistic and spiritual practices that might shed light on daily life in those prehistoric days. Mudit Trivedi, the anthropology graduate student who discovered the carvings, has recognized their impor-tance and has emphasized the need for further research on this and other archaeological sites around the campus.[14] But so far, his pleas have been ignored.

This reflects a more general neglect of Delhi's prehistoric past, which has not been a priority for government funding agencies. Meanwhile, the intensive, ongoing urbanization of the Delhi region has sent real estate prices sky-rocketing and has caused many archaeological relics to be bulldozed aside. It is no coincidence that significant archaeological finds have happened at construction sites in Delhi, from Anangpur to the JNU campus. But this is hardly a sustainable strategy for archaeological explorations. Construction crews turn up ancient relics only to bury them under layers of concrete; it's only through fortuitous interventions that archaeologists pluck some relics of note from the churning earth of Delhi.

Little attention is paid as the evidence of Delhi's most ancient settlements is destroyed piece by piece. By way of comparison, the destruction of archaeological remains in war-torn Iraq has grabbed international headlines, and has been used by various political commentators to point out the mindless violence of the US Army or ISIS, depending on the commentator's political persuasion. The destruction of Delhi's ancient archeology has been a quieter, more mundane affair. No war, just the humdrum expansion of a megacity—a mall here, a residential complex there, a factory here, a mining site there. It has upset the city's small coterie of archaeologists and history enthusiasts, but few others.

Even those who are passionate about the later archaeological remains—the massive fortress walls, the picturesque ruins, the exquisite tombs—seem indifferent or simply uninformed about the city's Stone Age past. In some ways, little pieces of chipped stone aren't as fascinating as intricately carved pillars and palaces, but they open a window into a much deeper, more mysterious past, one far removed from Hindu myths. Far from being an eternally Hindu city, Delhi is actually a symbol of a much broader humanity—a reminder of our African origins and our uncertain journeys into the unknown.

Stone Remnants of a Different Sort

Despite all my wanderings in the Ridge, I have yet to chance upon a Stone Age tool. Perhaps I don't have the attuned eye of the archeologist. Perhaps I am just unlucky. So I decide to set myself a more realistic target. On a sunny day in late November, a friend and I set off on my motorcycle to search for a historically significant bit of quartize that is, I have heard, less elusive than the Stone Age tools scattered across the city. It's a long journey from my home in Mehrauli, on the southwestern edge of the Ridge, to the neighborhood called East of Kailash, on the Ridge's southeastern side. Not that the two neighborhoods are so far apart geographically; the trip is long because of the notorious traffic of Delhi, a city that famously boasts more cars than Bangalore, Chennai and Mumbai combined.

We finally reach our destination: a small park maintained by the Delhi Development Authority (DDA). The park is a welcome relief from the congestion and aggression of the Delhi roads. Tibetan prayer flags are fluttering in the breeze. Groups of Buddhist monks, with shorn heads and purple robes, clamber over the rocks and pose for selfies. Two men are sitting on a bench, rolling a joint. Others are lying down in the grass, soaking in the winter sun.

This is Delhi at its most idyllic. Not too hot; not too cold; rolling hills and green grass; people enjoying picnics and kids running around. In some ways, the park is typical of the city. There are scores of other carefully manicured, fenced-off parks just like this through-out the metropolis. But the presence of the prayer flags and the monks suggests that there's something special about this one.

This suspicion is confirmed when we pass through the main gate of the park, skirting the garbage dump that borders the gate. A small signboard, with text in both English and Hindi, announces that this is the site of an Ashokan edict, discovered by the Archaeological Survey of Indian (ASI) in 1986.[15] Another late discovery!

Most Delhi-ites aren't aware of this piece of history in their midst, but it seems to be well known internationally. The monks visiting the edict are from Myanmar. Another group from Sri Lanka is on their way. They have come to pay homage to the edict from Ashoka, the great emperor who famously converted to Buddhism, and sought to instill Buddhist values throughout his empire, which covered vast stretches of the Indian subcontinent. Today, the Delhi edict is enclosed in a metal cage installed by the government, which is in turn protected by a concrete structure. A kindly caretaker, employed by the ASI, holds the keys to this structure, and opens it up when tourists approach. He tells us about the history of the site.

The edict has been carved into one of the quartzite rock outcrop-pings that dot this area. Ashoka wanted his legacy to be written in stone, quite literally. Though the Stone Age was long gone, stones still packed a symbolic punch.[16] His rock edicts, including the one in

Delhi, tell of his conversion to Buddhism, his great achievements, his exhortation that both the humble and the great follow the Buddhist path, and his hope that this cause would "endure forever". Embedding this message in stone—the painstaking work, no doubt, of a skilled craftsman—does give it a kind of permanence, but even that (as any good Buddhist would know) is illusory.

The text of the Delhi edict is now almost entirely worn away. This is in part because devout visitors insist on touching the stone, to establish some kind of tactile communion with this revered Buddhist ruler. But despite the near illegibility of the writing, the visiting monks do clearly see themselves as carrying on Ashoka's cause. A small statue of the Buddha has been installed near the edict. He is surrounded by wilting but colorful flowers. Incense burns, and visitors leave money: Indian rupees, American dollars, Burmese kyats. On their way out, many pick up a pebble from the surrounding stones as a keepsake.

The presence of an Ashokan edict in Delhi suggests that the region was important during the period of Mauryan rule, which was at its peak in the third century BCE. This area would have fallen on the Uttarapatha, the major trade route of the region, and would have been important ecologically because of its location at the center of the watershed formed by the Indus and Ganges rivers. Ashoka's reign is often presented as a golden age of Indian history, marked by religious tolerance and vibrant cultural expression, as well as material wealth. But apart from a fading stone inscription, little can be said about life in Delhi at this time. From the available evidence, it is unlikely that there was any kind of urban agglomeration in Delhi during this era.

Still, there is no doubt that the Delhi of the stone edict was radically different from the Delhi of the Stone Age tools. The existence of empires like Ashoka's reflects the full-fledged emergence of powerful states ruling over class-based societies. Huge empires were increasingly common, and with them, increasingly intensive practices of agriculture, which could satisfy the material demands of growing states.

At this point in history, and indeed for more than a millennium after Mauryan rule, Delhi was not seen as an ideal base for empire-building, largely because of its arid climate. It became more attractive to empire-builders only in the eleventh century CE, for reasons that will be explored in depth in the next chapter. With the region's rise to prominence, Delhi's stones found themselves unearthed for a new purpose: fortifying state power.

The first states to show interest in Delhi gravitated towards the Ridge, and the first stone fortresses were built in what is now Mehrauli. The initial walls were erected by a clan called the Tomars, who were defeated by the Chauhans, who were in turn defeated by Mohammad Ghori and Qutbuddin Aibak, who set up the first lasting Turkic Sultanate in Delhi (more on these characters in the next two chapters). This quick succession of rulers in Delhi points to the city's importance in an age of widespread military contestation, due to its position as a gateway into India from the subcontinent's northwest. Until the Mughals set up a relatively lasting peace, north India was roiled by waves of conflict, as powerful empires from Central Asia competed with each other and with local dynasties for control of the region.

In this tumultuous environment, it is no wonder that the early rulers of Delhi were largely concerned with securing their own safety. By locating their citadels on the rises of the Ridge, they provided themselves with a useful vantage point to spot marauding armies and withstand sieges. And they built enormous stone walls encircling their cities, making each settlement its own fortress.

Not only were these fortifications located on the Ridge, but they used the very stones of the Ridge to construct their massive walls and buildings, unearthing huge quartzite blocks to create their fortresses. In other words, the early urban settlements in Delhi were not only built on the Ridge, they were built *of* the Ridge. The ruins—especially the remains of a fort called Tughlaqabad, which tower over the adjoining road—are a bit like the Aravallis themselves: eroded reminders of previous days of glory, now battered by many years of decay. But the craftsmanship, and the attention to geological detail, is still evident. In the ruins of Lal Kot, there are quartzite blocks that fit together with perfect precision, thus making mortar unnecessary.

By the time of the Mughals, urban Delhi had shifted away from the Ridge and onto the banks of the Yamuna River, which was becoming increasingly important for trade and as a water source. In maps from

Mughal times, the Ridge has receded into the distance, providing a backdrop to the foreground of the thriving city. With the Mughals, the Ridge became the outskirts of the city, or, as in the case of Mehrauli, a romantic getaway, far from the responsibilities of state-building.

Quarry Quarrels

The British brought the Ridge back into prominence. While they engaged with the Ridge in myriad ways, adding layers of military, emotional and political significance to the site, they certainly did not forget its foundational rock-ness, and its value for constructing massive buildings.

In 1803, the British East India Company took control of Delhi and ruled by proxy through a series of Mughal kings who played increasingly ceremonial roles. It further tightened its grip on the city after brutally quashing the Uprising of 1857, an event which prompted East India Company rule to be replaced by the direct control of the British Crown. But the wholesale British transformation of Delhi only began in earnest in 1911, when it was named the new capital of British India, replacing Calcutta.

The construction of the gleaming, rigidly geometric New Delhi (the initial phase of which took roughly 20 years) brought an unprecedented building boom to the city, and the quartzite blocks of the Ridge provided suitable building blocks for government buildings, while also providing continuity with empires past. Even before that, as early as the 1870s, Delhi quartzite was mined by the British and used to construct structures like the Agra Canal, which diverted water from the Yamuna. But the building of New Delhi led to mining on a much larger scale.

Clues about this mining boom can be found in the Delhi State Archives, which sit facing the Ridge in south Delhi, not far from Mehrauli. Tucked in a quiet corner of the Qutab Institutional Area, the archives contain the records of the erstwhile Deputy Commissioner and Chief Commissioner, imperial posts for mid-level British officials charged with administering Delhi District (as it was designated at that time).

Many of the files in the Delhi State Archives are starting to disintegrate and others have disappeared altogether, but the administrators of the archives have kept alive the British spirit of bureaucracy and red-tapism. I discovered this first-hand during my many visits to the archive, which involved countless forms in triplicate, meticulously-maintained handwritten registers, and mysterious procedures that governed access to files. I often found myself gazing wistfully out the window of the archives' reading room, watching the swaying trees in the nearby Ridge park, as I waited for a particular form to be processed or a particular approval to be granted.

But it was worth the effort to get my hands on the weathered letters penned by British officials and to get a peek into the world of colonial governance. The archival material on Ridge quarrying is especially fascinating. These files reveal the government's obsession with control, regulations and bureaucratic processes. Even something as elemental as stone gets bureaucratized, and thus is subject to endless power struggles and jurisdictional battles. Quartzite, for the British, was largely a means to an end: a way to make the empire look grand, and, perhaps just as importantly, a way to exert financial control over natural resources.

A look into one particular set of archived correspondences reveals just what was at stake in the British efforts to mine the Ridge. India was described as the jewel in the crown of the British Empire (I hardly need to point out that this is a geological metaphor), and the new Indian capital needed to reflect this grandeur. At the same time, the government was obsessed with cutting costs, especially as critics at home in England accused India's British administrators of being irresponsible, corrupt, and profligate. British India, which was meant to bolster the Empire's economic dominance on a global scale, was seen by some as a liability. The planners of New Delhi, then, had to balance the need for regal splendor with the need for fiscal scrupulousness. As one analysis put it, "The maximum effect was to be obtained at a minimum of cost—the *leitmotiv* of British rule in India."[17] The mining of the Ridge must be seen in this context. Aravalli stone was abundant, locally available, and grand enough for the new capital. It is little surprise, then, that the Ridge was mined intensively as New Delhi was built.

Even with a constrained budget, this was a massive undertaking. In addition to Ridge rock, stones were brought in from other parts of northern India, most notably the pink sandstone that now adorns the most important buildings in New Delhi. The British built special rail lines to carry rock from the Ridge quarries and beyond directly to the building sites in New Delhi, and the stone yard created for this purpose was the biggest in the world at that time. The British brought in not just stone, but stone-cutters, casting a wide net across north India to find craftsmen skilled in shaping rock. A recent book about the making of New Delhi, emphasizing the speed and scope of construction, notes that the combination of skilled workers and modern mechanical devices like cranes "created an unprecedented momentum of construction work".[18]

But this momentum was difficult to sustain. In 1916, chaos erupted on the mining front, albeit in the stiff-upper-lip British bureaucratic fashion, in a dispute that drew in a wide range of powerful officials and their underlings. What started as a concern over sanitation turned into a larger turf war that raised crucial questions about government ownership of resources like stone and processes like mining in an imperial city. It's worthwhile going into the details of the case, since it shows that the imperial system, though dysfunctional at times, rarely lost sight of its key goal: to extract as much revenue as possible.

On 16 February 1916, the Chief Commissioner of Delhi, W. M. Hailey (after whom a road is now named in New Delhi), sent a letter to the Deputy Commissioner, noting that the deep pools of water created by quarrying were breeding grounds for disease, and suggesting that quarrying be limited to shallower depressions.[19] The Deputy Commissioner dutifully followed up on this suggestion, temporarily banning mining in the Jhandewalan and Panchkuian areas, where particularly deep quarries had been dug.

An Executive Engineer from the Public Works Department replied, not with concerns about sanitation, but with territorial antagonism. His letter states,

My contractors Milka Singh and Girdhari Lal, who have been quarrying Delhi Quartzite...during the past twelve months, have

reported to me that the Tahsildar...stopped their work and informed them that they would now have to get a pass from the Deputy Commissioner. Will you please inform me whether the Deputy Commissioner has received any authority over these quarries from the Imperial Delhi Committee?[20]

Some explanations are in order. A "*Tahsildar*" or *tehsildar*, as it is more commonly spelled (the British were known for their questionable spellings and constant mispronunciations) is the head of a tehsil, a small administrative unit that became prominent during British rule. By the early 1900s, the role of tehsildar was played by Indian officials, who were making up a larger and larger part of the Raj's administrative apparatus. The Executive Engineer's letter suggests the complexities of British rule, as Indian contractors working for British engineers came into conflict with Indian administrators following the orders of British government leaders, with all the parties involved claiming to be working towards the same goal on behalf of the Raj.

But if all the players were on the same team, they were clearly not on the same page. The engineer's letter reveals a tension that Delhi has felt at least since the establishment of New Delhi as the new center of British power: the struggle for control between the local city government and the larger powers that be (the British imperial state, then the Indian national government). As the capital city, Delhi has been closely guarded over by latter organizations, while the former groups chafe under the constraints placed on them in governing "their" city.

In the case of the 1916 quarry quarrel, the main opponents were the Imperial Delhi Commission, tasked by the imperial government with planning the new capital, and the office of the Deputy Commissioner, which oversaw many essential administrative tasks for the District of Delhi. The Imperial Delhi Commission was part of the larger Public Works Department, whose engineers bristled at any challenge to their technocratic control over the newly emerging city. The turf war plays out in exceedingly polite language, with each letter invariably ending, "I have the honour to be, Sir, your most obedient servant." But beneath this veneer of professional humility, the officials exchanged sharp barbs and jockeyed for position.

Administration,
Delhi District

Chief Commissioner Hailey

Deputy Commissioner Beadon

Tehsildar
Fazl-ud-din

Imperial
Delhi
Commission

Chief Engineer,
Public Works
Department

Executive
Engineer,
Public Works
Department

Contractors
Girdhari Lal, Milkha Singh

Conflicting Colonial Hierarchies

The Deputy Commissioner, Major Henry C. Beadon, emerged as the protagonist of this molehill-turned-mountain. Looking back at the history of the Delhi Ridge in colonial times, Beadon seems to be everywhere at once. Besides his role in the quarrying controversy, he was a key player in attempts to extend agricultural taxes, a long-winded contributor to debates about forest conservation, and a leader of government efforts to buy land for New Delhi.

Beadon was a tireless defender of bureaucratic procedure. In the quarrying case, he attracted the ire, not only of the planners of New Delhi, but also of his supposed ally in local governance, the Chief Commissioner. Beadon, a stickler for the rules, repeatedly cited government regulations to defend his actions, while his opponents appealed to the urgency of building the new imperial city and exasperatedly implored Beadon to stop gumming up the works.

After Beadon stopped the mining of deep quarries, the Public Works Department's Chief Engineer responded to concerns about sanitation with irrefutable geologic fact, based on the Ridge's history of weathering and erosion: "it is necessary to cut the rock to a great depth in order to obtain sound stone free from partial decomposition." He then complained that "the output of stone is stopped and work at the Secretariats will shortly be held up."[21] He finally struck a deal with the Chief Commissioner whereby deep quarries would be allowed, with the provision that they be filled up with rubble as soon as quarrying ceased.

The matter could have ended there, but Beadon kept stirring up trouble (or kept on simply doing his duties, as he no doubt would have put it). He gave orders to announce an auction for the stones that had been quarried by the contractors Milka Singh and Girdhari Lal, to the great dismay of the Public Works Department, whose engineers needed that stone to build New Delhi's government secretariats. The Chief Engineer wrote an indignant letter to Beadon, who reluctantly agreed to stop the auction.[22]

But the long-suffering Beadon did not accept defeat. Instead, he defended his actions in a series of letters. In a representative correspondence, this one to the Chief Engineer, he started off with: "Your letter has evidently been written under a misapprehension of existing facts and prescribed procedure." He made it clear that, according to the government regulations, the Deputy Commissioner (that is, Beadon himself) should act on behalf of the State as the owner of State lands. He was thus incensed that the Public Works Department had been issuing licenses for quarrying on State lands. He tersely noted that "passes issued by any one but myself have been issued without any authority at all."[23] He suggested that the best way forward was to change the regulations so that the Deputy Commissioner no longer had responsibility over the area controlled by the Public Works Department.

This suggestion was implemented by mid-May, but, unsurprisingly, the tensions did not end there. Accusations and recriminations were hurled about through June and July, as a *tehsildar* (who, at this late stage, is finally given a name: M. Fazl-ud-din) once again stopped quarrying activities on a site that was extracting stone for the secretariats. Beadon blamed some of these problems on the incompetent drawing of boundaries, claiming that it was unclear where his turf ended. He again claimed the moral high ground: "I have allowed the contractors to go on working in order to avoid delays: it will save correspondence if the... contractors are directed to comply with the published rules."[24]

Beadon also pointed to a deeper confusion about the limits of his power.

If the land is to be under my charge... for quarrying, it is essential that the land should be under my full and unfettered control. No Executive Engineer...should have anything to do with the permits for quarrying.... If a departure is to be made from the Regulations in respect to this matter, the most feasible situation will be for the whole...area...[to be] removed entirely from the Deputy Commissioner's control.[25]

Beadon's desire for "full and unfettered control" set the stage for many future debates over jurisdiction and city planning. Frustrated with the conflicting demands of different departments, he believed that each parcel of land should be under the clear, direct control of only one agency. It is not power he wanted, so much as clarity and lack of interference. He was happy to give up control of New Delhi lands; just don't give him partial control.

In the same letter, Beadon explained why he thought this exclusive control was necessary. Mustering his argumentative forces, Beadon cited the Land Acquisition Manual, the Settlement Manual, and a Standing Order of the Financial Commissioner, all of which showed "very clearly that royalty is a form of land revenue and that it has always been the duty of the Deputy Commissioner to collect royalty and to issue passes for quarrying on all land." Criticizing the Chief Commissioner's approach, Beadon asserted that his suggestion to hand lands over to the Public Works Department

> *was accepted...but in a modified form, which I notice does not provide for the levying of royalty or land revenue when quarrying takes place in those areas. Presumably private individuals and contractors will (if they are not at present) practice quarrying there. Unless the responsible officers start a proper system of permits, the State will lose its land revenue.*[26]

With this, Beadon struck at the heart of the matter. The imperial government was nothing if not a revenue-collecting machine. As a colonial instrument, its main function was to extract resources (natural, human-made, financial) from India in service of the empire. Beadon's claim is that streamlined, clear-cut administration is essential for revenue collection. Stone, then, was an important building block for New Delhi, not only in a literal sense, but in a financial sense. The government's ability to levy royalties on quarried land was a revenue source that contributed, however modestly, to the notoriously expensive construction of the new imperial city.

Beadon's pleas fell on deaf ears; or at least, no responses are recorded in the archives. He was outranked, after all, by his opponents, within both the local and the imperial governments. But Beadon's lesson—that government control over geological resources means government control over revenue—was hardly lost on his compatriots. Nor was it forgotten by those who took control when the British Crown left India.

The Life and Death of Mining in Postcolonial Delhi

After Independence, the new government, like the old, attempted to exert control over the quarrying of Ridge stone, although they had to confront increasingly powerful local interests who wanted their own share of the mining wealth. There is evidence of quarrying activities on many sites throughout the Ridge in the years immediately following Independence, but by the 1960s, the stone industry had centered itself firmly on the southern border of Delhi, more than fifteen kilometers away (as the crow flies) from the imposing imperial monuments of New Delhi that had been appropriated by the country's new postcolonial leaders.

What emerged in the 1960s and 1970s was a village-based quarrying industry dominated by local elites, who threw open Ridge land to feed the capital's ever-growing construction industry.[27] The far south of Delhi, although within the boundaries of the capital, was still entirely rural in character, and small villages dotted the southern border, from Badarpur in the east to Kapasera in the west. Panchayat leaders began to see the hilly parts of village commons—which had previously been used collectively for grazing or forestry—as potentially lucrative investment opportunities, and a thriving mining industry soon developed.

Quartzite blocks remained an important product from these mines, but another component of Ridge geology also rose in prominence. While the British emphasized the importance of digging deep to find uneroded stone, postcolonial miners found that there was great value precisely in the Ridge stone that was most eroded, since it proved indispensable for the next stage of Delhi's growth. This material too is derived from iron-rich quartzite, but in its most weathered, broken-down form. The outermost layer of highly weathered Ridge quartzite eventually breaks down into a thick reddish sand. Locally, this is known as Badarpur sand, named after an area in southeast Delhi where the sand is particularly prominent. Badarpur sand is often used as an ingredient in concrete, and it is also useful for filling in the foundations of buildings and for constructing roads. As the city of Delhi has been increasingly hemmed in by concrete and pavement, Badarpur sand has played a crucial material role.

Mining in the post-Independence period was both more extensive and more thoroughly documented than colonial-era mining. It
extended throughout the Aravallis, but was particularly intensive in
Delhi given the rapid pace of urbanization and the resulting need for
construction material. In some ways, not much had changed since
British times. The government was still the ultimate authority when
it came to mining; any land that was quarried technically came under
government ownership (a nation-wide practice that was only questioned in 2013, after a surge of reforms in the mining industry), and
it was the Delhi government's responsibility to issue permits. Like
the British government before it, the Delhi government—which, until
1991, was directly administered by the central government—was keen
on maximizing revenue. One way they did this was by auctioning off
leases for quarrying on Ridge land.

But this is just the beginning of the story. The official winners of the
auction—the lessees of quarry-able Ridge land—were often panchayat
members who already enjoyed social and economic privileges within
the village hierarchy; various efforts by workers' collectives to win
quarry auctions were stymied by their lack of funds and by their
opponents' political maneuvering. The typical lessee, while locally
powerful, had little involvement in the day-to-day operations of
the quarry. He served mostly as a figurehead who received various
economic and political benefits from the role.

The actual work was overseen by a group of contractors and middlemen, referred to in local parlance as *thekedars*. Backed by local
musclemen, if not by any legal claims to the quarries, *thekedars*
marked out their spheres of influence and maintained exclusive
control over their domains. Below the *thekedars* were another essential link in the chain of control: *jamadars*, or labor contractors, who
found groups of workers with the requisite quarrying skills. At the
bottom of the pyramid, in terms of both income and social standing,
were the scores of workers. Even the workers were subdivided and
categorized, with those who cleared vegetation and debris from the
rocks ranked below the workers who did the actual quarrying.

The workers were largely migrants or children of migrants from
Rajasthan, with a smaller percentage coming from Haryana and
Uttar Pradesh. Most often, they migrated to the Delhi mines after

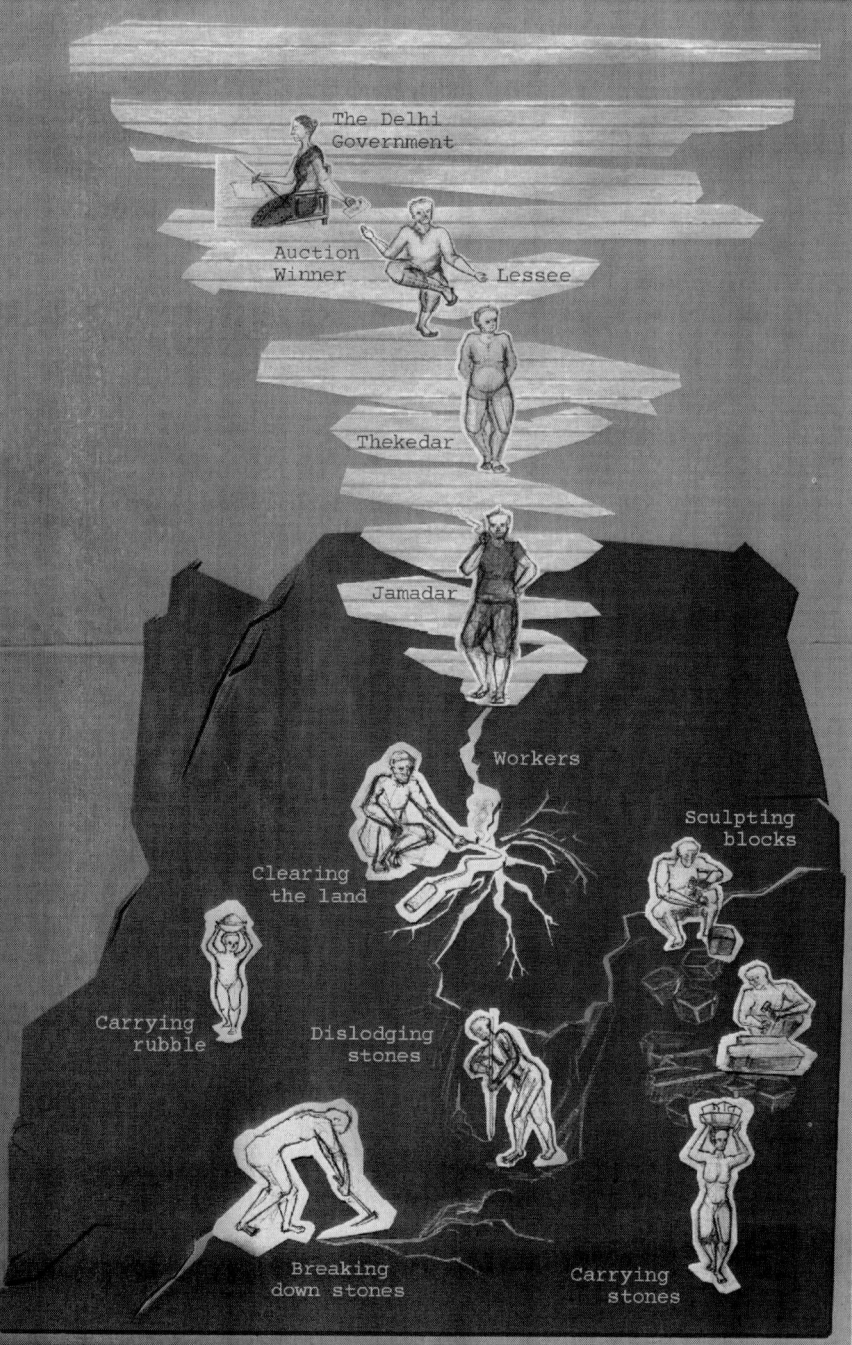

Hierarchies in the Quarries

drawing on the financial help and social connections of a *jamadar* who was originally from their region. They brought with them vivid memories of agricultural decline and landlessness in their rural homes, and oppression from the upper-caste elite in their villages (the stone workers were invariably "lower" caste). They came to Delhi more out of desperation than out of hope for a transformed life. In the apt words of Mohammad Talib, the *jamadars* represented "a case of the city reaching out to the village to recruit the prospective proletariat".[28]

Of all the contemporary chroniclers of Delhi's stone quarry workers, Talib has probed most deeply into their material conditions, their social and economic life, and their strategies, dreams and coping mechanisms. His outstanding ethnographic account, *Writing Labour*, draws on three years of intense interaction with the workers (from 1984 to 1987), as well as extensive textual research.

Talib does not mince words about the highly exploitative nature of the quarrying business, although his condemnation is couched in academic language: "Stone quarrying as a business venture involves a minimum of capital investment in the actual work and its administration against a maximum of profiteering and capital augmentation."[29] In other words: the lessees and *thekedars*, once they secured mining rights from the government, needed to invest very little money while making extremely high profits due to unethical and exploitative business practices.

Meanwhile, the stone workers risked their life and their health for extremely low pay, and even had to provide the tools for their own work. Even worse, the workers often had to take loans from *thekedars*, *jamadars*, and other unscrupulous lenders to collect the money to buy tools and other essentials, including food that contractors sold at inflated prices. This often led to a spiral of debt and the workers' complete dependence on their employers, which then resulted in conditions of bonded labor: workers no longer received a salary, and worked only to repay their ever-mounting debts.

Talib also provides extensive details about the excruciating labor done by the stone workers. The initial work of clearing the land to be quarried was done by the lowest group on the workers' hierarchy, members of a caste that has traditionally used donkeys to transport materials. In this case, though, the workers did their own carrying, with men typically digging up the soil with a pickaxe and woman or children (who were inevitably paid less) carrying the earth in baskets to a convenient dumping ground. According to Talib's calculations, over the course of a day, a worker would carry about 8,000 kilograms of earth and would earn around ₹9.

Then came the actual quarrying work, carried out by another set of workers who were slightly higher in the hierarchy. If possible, these workers removed stones manually, so they did not have to use expensive, dangerous blasting equipment that they themselves had to buy. After digging down until they hit a vertical rock face, the workers then used iron rods to pry loose large chunks of stone. If this was not possible, out came the explosives. Experienced workers were able to intuit where to place the explosives to get the most effective blast. These experienced workers were also in charge of determining the length of the wick and lighting it with a *beedi*, before all the workers retreat to a sheltered spot.

Once the stones had been detached from the rock face, the next step involved breaking the stones into convenient pieces. Much of the stone was broken into irregular pieces, but the most valuable products were stones shaped into standard-sized chunks, which were then used for milestones or for bricks. Breaking these stones took incredible strength and precision, and were thus the provenance of another set of workers. Under a master-apprentice system, these workers generally took about three years to develop the ability to make near-identical bricks by eye, without any measuring tools. At most, the finished products varied by a couple of centimeters. Finally, yet another set of workers, lower down on the scale, carried away the broken-up stones and load them into trucks.[30]

While Talib's attention to the grueling, exploitative nature of this work is commendable, what gives his account unique depth is his analysis of the work's larger symbolic importance. The workers depended on quarrying for their livelihood, and thus the stones became an important part of their symbolic universe. The process of laboring in the stone quarries clearly transformed the workers, even as they transformed the landscape and extracted the raw materials for Delhi's booming construction industry.

As the workers sought to make sense of these transformations, they often resorted to geological similes or metaphors. Talib reports, "One worker actually showed me his palm with a claim that it was as hard as a stone and would remain unhurt even if a truck rolled over it. The worker was suggesting that working on stones turned his hands into stones."[31]

The workers' metaphorical invocations of stone should not be taken lightly. These were not just literary flights of fancy; rather, such expressions were grounded in their intimate relationship with the stones that ensured their livelihood. Their day-to-day interactions with the stones formed a vital framework for seeing and understanding their world. One worker remarked, "We were born amidst stones, live in them, and get our bread from them."[32] As is clear from many accounts of the work itself, this was a matter of life and death. Not knowing the stone well enough—how explosives will fracture it, how pieces of it will fall from a rock face, how easily it will break—could result (and often did result) in horrific injuries and even death. Adding to these practical concerns was the sheer amount of time the workers spent around rocks. This often began in childhood, as many of the workers grew up around the quarries, and spent their younger days helping their parents, who were themselves quarry workers. The children often spent many hours collecting stones, informal training for their eventual occupations.

The quarry workers thus used stones as source material for rich conceptual constructs, filtering their other experiences and struggles through the hard reality of the material. One of the main uses of stone metaphors was simply to express the strenuous, physically taxing nature of the job; Talib quotes a working saying, "Breaking rocks requires an unbreakable body and mind."[33] As the workers got older, the metaphor became considerably darker: "Our bodies have become as the pits in the quarries."[34]

Beyond this relatively straightforward equivalence, stone metaphors were used much more expansively as workers sought to explain their moral universe. Several workers employed these metaphors to assert the importance of patience and equanimity: "A hard rock tells us to remain patient and never to lose heart." Similarly, "You cannot break rock without breaking your anger. Quarry work is never done alone; you work with your fellow workers. If you are quarrelsome you lose them and your bread too."[35] And in a more resigned vein: "To complain is to break one's head against a rock."[36]

These attempts to come to peace with the grueling work sometimes led to a muted acceptance, with some even using the language of occupation-based caste to assert the inevitability of such work: "Hands that deal in stones must not deal in any other material. Different things require a different aptitude to handle."[37] Yet, on other occasions, and for different workers, similar metaphors were used more subversively, to demand more equitable working conditions: "Our work is as hard as the stones. Our wages must match our hard work."[38]

Throughout the 1970s and into the early 1980s, the workers mounted increasingly ambitious attempts to unionize and collectivize. There were many collective actions, from workers' cooperatives to lengthy strikes, at times supported by civil society organizations, at times relying on local leadership. The workers not only had to confront the contractors, mine owners, and a growing set of middlemen, but also a paternalistic state that repeatedly affirmed its support for the workers' rights and their basic dignity, while quietly letting labor abuses go on unchecked and striking backroom deals with contractors. For each worker victory, there was an even harder pushback. But throughout, the workers' language of resistance, just as much as their language of quiescence, was laced with stony metaphors.[39]

Ultimately, though, the question of resistance versus quiescence ceased to matter. In 1983, a directive from the Delhi government banned quarrying activity in the city. The decision was ostensibly for the sake of workers, who had never given up the fight for better working conditions, even as the state of the mines deteriorated. The frequent injuries and gruesome deaths were an embarrassment for the government, which played an increasingly large role in the management of the mines from the mid-1970s onwards. Banning mining solved a bad PR issue However, the ban was only sporadically enforced, a negligence made painfully clear by the death of seven workers in a pit-side collapse in 1990. A more wide-ranging ban was finally put in place in 1991.

Though the closure of quarries in Delhi had the supposed aim of protecting workers, this was cold comfort to the 4,000 people who suddenly found themselves out of jobs. They must have been quite surprised to hear the state's rhetoric of "worker's safety"; after all, this was the same government that had "overseen" the quarrying industry in Delhi, and through a combination of malign neglect and active collusion with powerful village leaders and an intricate web of intermediaries, had been responsible for the exploitative and unsafe working conditions of the quarry workers. The workers protested that the mines should be re-opened, with stronger safety measures in place. However, the Delhi government found it easier to sweep the entire issue under the rug. Further, the quarries of Delhi were offering diminishing returns, both geologically and financially, as quarrying crews were starting to descend to groundwater levels, at which, work is no longer feasible. Banning mining opened up other alluring possibilities for the use of the land, such as the development of high-end real estate. Of course, mining continued, but covertly, or else across the border in other parts of the Aravallis.

The closure of the Delhi quarries coincided with a rise in the popularity and the power of environmentalist rhetoric, a trend that will be dissected much more thoroughly in later chapters. The largest of the quarrying sites, the Bhatti mines near Badarpur, was incorporated into a nearby forest reserve; thus the Asola Wildlife Sanctuary became the Asola-Bhatti Wildlife Sanctuary.[40] Yet, while some government officials were genuinely enthusiastic about environmental conservation, the use of an ecological justification for the closures acted largely as a smokescreen, obscuring other motives. This became increasingly clear in the years after the closure, as ecologically harmful, non-native, fast-growing species were planted to maximize the appearance of a green cover in some parts of the Sanctuary, while trucks carrying stones from illegal quarrying activities continued to ply the roads of the Sanctuary.

Worse, in other former mining areas, the land quickly developed into sites for much-coveted, high-end luxury estates, especially the misleadingly named, palatial "farmhouses". Urban Delhi was rapidly expanding, both geographically and economically, and the former rural "hinterland" areas were becoming increasingly desirable. The conversion of mines into luxury housing was no accident. As the fiercely committed activist and journalist Anita Soni notes,

*It would be wrong to assume that the termination of quarrying...
has been the cause behind spectacular increase in the number of
farmhouses and in the acreage of existing estates. In reality, the
deals had been struck much in advance, during the years of the
operation of the quarries.... [The quarrying ban] was expected,
and awaited, by the village-based intermediaries, ready to
exercise their ownership rights over the land.*[41]

Some former quarry workers found employment performing
menial tasks on the sprawling estates that sprang up in this zone.

The ban on mining thus appears as a kind of magic trick, making
quarries disappear and replacing them with eco-friendly sanctuaries.
Of course, like any magic trick, this is an illusion. Quarrying still goes
on, because its products are essential to the ever-growing construction industry; it just happens outside of city limits. Actual quarrying work is messy: it involves real people doing real labor, oppressed
by real exploitation and organizing real protests and real collective
actions. Much better to cover this up as much as possible and replace
it with a simulacrum of pristine nature. And some forest-land *is*
actually created, but it is not what it seems, and it is hemmed in by
ever-growing mansions.

This sleight of hand in the Delhi Ridge mirrors later developments
regarding the Aravalli Range as a whole. In 2009, the Supreme Court
banned mining in the entire Aravalli range, citing, of course, ecological
concerns, including the erosion of water courses and the loss of green
cover. But the construction industry needs its raw materials, and
Delhi continues to grow, which ensures that mining will continue in
one way or another.

In the Supreme Court judgment, there was no mention of the people
who actually work in the quarries, and depend on them for their livelihood. Like the Delhi government, the Supreme Court has seemingly
forgotten that these people exist. Their work, and their intimate relationship with the Aravalli stone, has been relegated to the shadows.

Touch the Rock

The modern-day history of mining and mining bans in the
Delhi Ridge brings to light the broader pitfalls of anti-ecological

environmentalism, which prioritizes pleasant scenery close to home, but fails to examine the roots of environmental problems, and our own potential complicity with these problems. We want nice parks and lush forests, but we also want inexpensive building material for ever-bigger homes. Without quarrying, there is no concrete for high-rises, there is no asphalt for roads. But thanks to mining bans, these quarrying activities are out of sight, and thus out of mind.

If the stone of the Ridge exists in the consciousness of the city's elite, it has lost its economic weight and has become a breezy site for recreation. Rock climbing has become an increasingly popular pastime in the national capital, with the prime outdoor climbing sites located, of course, on the Ridge. Instead of workers toiling in the quarries, there are climbers following the mantra, "be one with the rock". The stones are no longer a means of sustaining life; rather, they are a way for the relatively well-off to escape the drudgeries of urban existence.

The history and the socioeconomic overtones of rock climbing are quite complex. Mountaineering, the more general pastime that gave birth to rock climbing, has a long history as an elite, hyper-masculine pursuit, strictly regulated along class lines by etiquette-heavy organizations like the Alpine Club in the United Kingdom.[42] In the 1960s, in keeping with the mood of the times, mountaineering got a more free-spirited, counter-cultural edge, as well as an increased awareness of environmental issues. It was during this period that modern rock climbing emerged, with its emphasis on a minimalist style that uses limited amounts of gear and strives to avoid damaging rocks and leaving debris behind on climbs.

American rock-climbing pioneers like Yvon Chouinard came from working-class backgrounds and embraced a vagabond lifestyle. Spending their nights in campgrounds near climbing sites, they hunted squirrels for dinner, and made money by designing and creating a new generation of eco-friendly climbing tools, setting up makeshift factories in their campsites. Now Chouinard is the founder of the wildly successful outdoor equipment company Patagonia, and he has appeared on the cover of *Fortune* magazine, which makes it difficult to argue that he represents a radical departure from climbing's history of privilege. But in his early days, he embodied the

2 SOIL
Mobile Ecologies, Hybrid Histories

Stony Soil, Thorny Trees: The Ridge's Unforgiving Ecology

On a bleak mid-January afternoon, the trees of the Ridge are draped in a thick, polluted haze. The sun shines weakly through the smog, but I still work up a sweat as I climb up a steep quartzite slope. A friend and I are in a park called Sanjay Van, just north of Mehrauli, Delhi's historic first city. The dirt trail we're following is more stone than soil; bits of rock and coarse red sand crunch underneath our feet as we make our way uphill.

We reach the top of the hill and take in the view. To the right looms the graceful tower of Qutb Minar; to the left are the ruins of Delhi's first walled fortress; in front of us is a sea of green. Behind us, the slope gently descends into a wide basin, remarkably flat in this hilly terrain. Years ago, some enterprising young men cleared all the trees and shrubs to create a grassy cricket ground. At the moment, two cricket games are taking place side by side, one for kids and one for adults.

As the cricketers laugh, lounge and play, a thin woman in a simple sari emerges from the forest and stands at the edge of the clearing, a huge pile of firewood balanced on her head. She is soon joined by a young boy and girl, each carrying smaller bundles of wood. They throw down their loads and sit for a few minutes, watching the revelry around them. Then they get up, and the two kids help the woman heave the firewood back onto her head. They skirt the edge of the cricket games and disappear into the forest on the other side.

Meanwhile, on the other side of the slope, we see a grizzled old man wrapped in a shawl, leading several goats through the underbrush. My friend calls out to him, and he replies gruffly. He is busy finding a suitable grazing ground for his flock, and he has little interest in the two odd figures yelling at him from the hilltop.

As we descend from our perch and head towards the park exit, we see more women collecting firewood, handling the branches carefully to avoid their prickly thorns. Throughout the year, firewood is a valuable fuel source for those who can't afford gas connections. But firewood is an even more precious commodity in mid-winter, when families make small fires outside their homes to beat back the damp chill of the season.

Near the exit of the park, we spot an official from the Delhi Forest Department, giving instructions to a group of contract workers about pending maintenance tasks. We chat briefly with the official, and as we exchange pleasantries, the group of women with their firewood crosses our path, heading home with their spoils. The official looks at them, looks at us, and lightly chides the women. They continue walking, unconcerned.

Technically, the women are breaking the law. So too is the old man grazing his goats. Sanjay Van is part of the Ridge's Reserved Forest zone, and as such is entitled to the state's most stringent protections. Firewood collecting and grazing are strictly prohibited, as an imposing sign at the park entrance makes clear. And yet, in reality, the prohibition is not so strict. The wood-collectors and shepherds have clearly come to an understanding with the Forest Department officials.

Some might dismiss this as a simple case of government corruption, but the truth is much more complicated, especially when one delves into the history of the area. Sanjay Van was named a Reserved Forest in the 1990s. For many years before that, it served as a commons for the surrounding villages, and it was used extensively for grazing and wood-collecting. These erstwhile rural areas have now been swallowed up by the expanding city (they bear the strange administration designation "urban villages"), but vestiges of an earlier life remain, especially for the area's poorer residents, who depend on the natural resources of the park. Elders in the area, who remember when their village truly was a village, resent the intrusion of the state and the restrictions on what was once their common property.

This dynamic is not just confined to Sanjay Van. Before the official label of "Reserved Forest" was imposed, the Ridge's primary ecological role—at least as far as humans were concerned—was to serve as a grazing ground and woodlot. This role was largely determined by the Ridge's geology, albeit in a paradoxical way. Although the Ridge's rocky soil is a harsh habitat for shrubs and trees, it was nonetheless the preferred location for gathering fuel and fodder.

To understand why, it's necessary to look at Delhi's ecology, and particularly its soil composition. The ancient Aravallis cut through the landscape of Delhi, leaving a band of rocky, infertile soil. Surrounding this is an environment shaped by a much newer and more imposing geological formation: the Himalayas, from whose glaciers the Yamuna river originates. The (relatively) new alluvium from the Yamuna river system gives the non-Ridge soil of Delhi its fertility. In traditional terms used for Delhi's soil types, the hilly Ridge land (called *kohi*) was distinguished from three different soil types that are broadly alluvial: *bangar* (fertile soil on level land), *khadar* (the sandy riverain of the Yamuna) and *dabar* (low-lying land that was subject to seasonal flooding).[1]

As human settlements in Delhi grew, they needed an ever-increasing food supply, and any zone with fertile soil was taken over by agriculture, which became increasing intensive over time. The Ridge, with its rocky soil, was not part of this agricultural zone and could thus be put to different uses: grazing, for instance, or firewood collecting. In fact, the stunted trees and grasses that managed to grow on the Ridge provided ample fodder for livestock, making it a prime location for pastoralists.

On the local scale, then, the Ridge's rocky soil serves to separate Delhi's pastoral zone from its agricultural zone. But this small-scale picture should be complemented by a much broader view. On a regional level, Delhi and its Ridge mark a transition from the arid habitats to its west to the fertile ones to its east. Thus, on both a microcosmic and a macrocosmic level, the Ridge serves as a dividing line between two kinds of soil and two kinds of ecologies, which in turn favor two distinct livelihood strategies—pastoralism and agriculture. In practice, of course, the distinction is not so neat. Agriculture and pastoralism are often practised in tandem, sometimes by the same people. The fact that Delhi has long supported both livelihood pursuits suggest that the city, and the region as a whole, represent a mosaic of ecological and economic landscapes.

along the southeastern edge of the Ridge. He needed more people to do the hard work of mining and hauling quartzite for the ramparts of his new metropolis, so he sent his officials to get workers from the old city. There, workers were building a step-well for the revered Sufi saint Nizamuddin. They reluctantly joined the imperial building project but hurried back every night to work on the saint's well.

Incensed, the emperor banned the sale of oil in the city so that workers could not light the lamps that illuminated their night work. Nizamuddin took water from the well itself and asked the workers to light the water as if it were oil; miraculously (for Nizamuddin was a miracle-worker), the water lit on fire. But the emperor had to pay for his insolence; Nizamuddin uttered a famous curse: "May your new city be inhabited only by jackals and Gujjars." For that sage, Gujjars were simply shorthand for wildness and barbarity. The prophesy was fulfilled shortly after Ghiyasuddin Tughlaq's death, as his son abandoned the new city less then a decade after its founding, likely due to severe water shortages in the area.

The marginalization of Gujjars continued with the coming of the Mughals, who seized control of Delhi as the Sultanate disintegrated. Babur, founder of the Mughal empire, famously complained about Gujjars and other pastoral groups who raided his cattle as he passed through their territory; the rear guard of his encampment chased down and beheaded several Gujjars raiders.[13]

The relationship between Mughals and Gujjars became especially strained with the founding of Shahjahanabad, the Mughal's Delhi capital, in 1639. According to Gujjar historian Rahul Khari, Gujjars inhabited the area where Shahjahanabad's Jama Masjid was slated to be built, a small hillock called Bhojla Pahari. Though this hillock does not figure in standard accounts of the Ridge, it deserves inclusion as a Ridge area. Such isolated hillocks appear on either side of the main Ridge formations in Delhi, and they are evidence of the city's basic geological and ecological layout. The ancient, much-eroded Aravallis are covered in many places by much newer alluvium from the Yamuna river system but pop out in places like Bhojla Pahari. These unexpected elevations became favored spots for religious monuments. It is no accident that the Mughals sited their most important mosque on one of Shahjahanabad's very few hills.

The Gujjars, no strangers to the hills of the Ridge, were, Khari tells us, not willing to give up Bhojla Pahari without a fight. Mughal troops went on the offensive. Gujjars from 12 surrounding villages were displaced in the violence, and were only able to re-establish three villages, one of which, Chandrawal, went on to play an important role in the colonial era.[14] The Mughals then leveled the top of Bhojla Pahari to create a solid platform for the mosque; this was a foreshadowing of the much more thoroughgoing leveling of Ridge land that would happen under later powers.

Given Khari's tendency to see Gujjar history through an extremely polarized religious lens, it is difficult to take this story at face value. Nonetheless, it reflects the undeniable violence and marginalization that Gujjars have faced at the hands of imperious city-builders in Delhi, including both the British and the postcolonial Indian state. If the account is not history, it is surely prophesy. It indicates the very real ways in which Gujjars have been written out of the history of the city.[15]

And yet, the relationship between Gujjars and the city of Shahjah-anabad was not one of unremitting conflict. Gujjar communities played an essential role in the economy of the city, taking advantage of the (relatively) stable urban growth that Mughal rule facilitated. As the city's population grew, so too did the demand for meat and dairy products, many of which were provided by Gujjar communities grazing their animals on the nearby Ridge. In addition to raising their own livestock, they also grazed the livestock of city butchers, taking them from the city to pastoral zones like the Ridge. In exchange, they were allowed to keep a handful of the animals, increasing their own flock.[16]

Gujjars and other pastoralists maintained part of the Ridge as grassy grazing land, while preserving other parts as woodlands, often in the form of sacred groves. Both forest and fodder zones were held in common, either by single villages, or by larger collectives that had developed customs and procedures for shared land use. For instance, in many places, Gujjar shepherds would pay a pasturage "toll" in areas that they moved through.

Such systems were starting to feel significant strain due to the continuing growth of Shahjahanabad, especially as the eighteenth century progressed. Landowners in surrounding villages started to complain about the pressure on their commons due to all the cattle that Gujjars were bringing in from the city.[17] It is likely that several areas in the Ridge became totally denuded during Mughal rule due to the extensive pressures on the land. But there was no question of overturning the system of shared land use and common ownership. This system, despite increasing urban pressure, still maintained large areas of both grasslands and woodlands on and around the Ridge, especially in the southern zones farther away from Shahjahanabad. And it was precisely this system, and the ecological pattern it maintained, that was irrevocably changed under British rule.

The British and a Radical Break with the Past

Ironically, it is only because of the British that we have a clear sense of the ecological backdrop of Shahjahanabad. The British took control of Delhi in 1803, and they surveyed their new territory with keen interest, recording their observations in various government reports and gazetteers.

One clear result of their investigation was the care with which communities maintained stands of trees. In typical dry prose, a gazetteer for Delhi District (an administrative unit created by the British) intoned:

Another characteristic incident of land tenure in the district is the reservation of wood-producing land in the shamilat deh [village commons] as an enclosure whence no fuel or wood is to be cut. This is generally connected with religion in the shape of a fakir's hut, or grave or a religious shrine; but sometimes no such religious element is observable.

The author of the report cannot help but add a note of imperial condescension: "The people, with that faculty of docile obedience which is at once such a help and a trouble (when it degenerates, as so often is the case, into slavish adherence to custom) to the administrator, observe the social precept without asking more about it."[18]

The administrators in question were the British, or more precisely, the East India Company. And they were not surveying the land out of mere academic curiosity. They were intent on making money from it. The East India Company was a strange vehicle of colonial rule, an amalgam of a money-minded corporation and an occupying force. It was, technically, "a monopoly joint stock company", chartered by the British Crown in 1600 to engage in trade, form local alliances, and compete with the rival monopolies of the Portuguese, Dutch and French. The British had little influence in India until the 1700s, and their rise was marked by rampant corruption, cheating and looting. Agents of the Company used its lucrative monopoly to line their own pockets, while leaving the Company itself in shaky financial straits.[19]

The British government eventually recognized the magnitude of this plundering, with its ill effects on royal coffers, and tried, with some success, to reel it in. But by this point, the East India Company, seeing the vast wealth of India, had ambitions that went well beyond trade. Through a series of deliberately misunderstood grants and treaties, along with military incursions backed by teams of mercenaries, the Company began piecing together an empire. Most histories of the British Raj point to the 1757 Battle of Plassey, when East India Company troops defeated the Mughal governor of Bengal, as the beginning of British territorial expansion. But an equally crucial turning point was the "Diwani" grant of 1765, in which the Mughal emperor gave the Company the right to collect agricultural revenue in Bengal. Through a questionable and self-serving interpretation of this grant, the Company reckoned that it now had the right to occupy the land as an imperial power.

The Diwani grant also points to a telling economic shift in Company policy; though trade was still profitable, Company officials had their eyes on an even bigger goldmine: land revenue. Agriculture was the basis of the Indian economy, and taxing agricultural land was thus crucial to funding Company rule. This became increasingly true as the British expanded beyond their initial coastal footholds. Thus, when the British took control of the Delhi area, one of their priorities was to assess the productivity of the land and find an efficient way to tax the population based on their assessment.

Looking out on their new lands, British administrators were struck by the preponderance of "waste" in Delhi District.[20] For the British, "waste" had a clear economic definition: if land was not

Touch the Rock

rebellious spirit of a community that chose to live in the woods and on the rock-face instead of climbing the career ladder.

Like many activities imported from the West, rock climbing has lost this cultural complexity and has entered India purely as a pursuit of the well-off (see also: bowling, long associated with the American working class). Climbing gear, like the industry-standard gear made by the other company Chouinard founded, Black Diamond, must be imported from other countries at prohibitively high prices. The vast majority of Delhi climbers are "weekend warriors", taking time off from their busy jobs to unwind and challenge themselves with an unusual, energizing pastime.

Many of these climbers make use of the "Delhi Rock" Facebook page, which serves as a popular online meeting place for outdoors enthusiasts living in or passing through Delhi. The administrators of the page post information about climbing spots in Delhi and welcome newcomers to the climbing community.

The Delhi Rock group also organizes climbing trips to two main locations: the so-called "Old Rocks" of Lado Sarai, in a Ridge park in the midst of urban South Delhi, and Dhauj, a village well outside of Delhi city limits. While Lado Sarai was once its own village, it was long ago engulfed by the rapidly expanding city. The park surrounding the "Old Rocks" is well maintained by the Delhi Development Authority, with wide dirt paths and benches sprinkled throughout. Delhi Tourism, a government body, even tried to promote the park as a rock-climbing site, painting numbers on the rocks to label various routes, and installing a sign with stern rules. It never quite caught on, and now the sign is rusted to the point of illegibility. Still, groups of climbers descend upon the site from time to time, joining the rowdy cricket players, the old men conversing intently on benches, and the local athletes there for a morning jog or workout.

Dhauj is more remote. It is still a village, though given its proximity to Delhi, it seems likely that it too will soon be swallowed up in the sprawl of the megacity's suburbs. A few canny businessmen have tried to cash in on Dhauj's relative quiet and its natural beauty; the area now boasts two complexes, "Camp Dhauj" and "Camp Wild", that offer comfortable accommodations and host "adventure" activities like rock climbing, rappelling, and mountain biking. Reports on the "Delhi Rock" Facebook page, though, suggest that all is not well at Dhauj, and that the villagers resent the intrusion of city folk, including the many foreigners who have joined Delhi Rock trips.

A recent post on the page recounts a harrowing incident. A group of climbers had just completed a climb, and were about to start another, when they saw that their rope had been cut with an ax by a local teen. Perhaps the teen did not recognize exactly what he was doing, but his act of vandalism could have been deadly; if the climbers had not noticed the cut rope, and had continued climbing, a fall would have proven fatal. Perhaps the teen was feeling malevolent; perhaps he justified his action because it occurred on his village's ancestral grazing ground, and he was feeling protective. Whatever the case, the incident reveals the ugly side of a culture clash, pitting an adventuresome elite and a foreign activity against a village that is being inexorably drawn into the urban sphere.

The dynamics of Dhauj place urban climbers and rural herders on opposite sides of a stark divide. But the climbers bear some remarkable similarities to another marginalized group: the stone quarry workers that once populated Delhi's margins. Both groups take significant risks while interacting intimately with Delhi's stones; both rely on their first-hand, sensual geological knowledge to keep themselves safe in precarious situations; both come to see the rock as part of their symbolic world. But of course, this overlooks their fundamental differences. The workers were pushed into these jobs by the force of circumstances, while rock climbers freely choose to experience these risks. For those mining the quarry, it is inevitable, regrettable work; for the climbers, it's an escape from work.

If it weren't so unjust, there would be something poetic about the replacement of workers with climbers on the Delhi Ridge. Certainly, it's a sign of a certain vision of the city, one that wishes to push aside the messiness of work (and working-class politics), to be replaced by scenic, tension-free sites of leisure. And this vision is hardly unique to Delhi. The artist and cultural critic Martha Rosler has spoken of the "Disneyland" model of the city, which replaces class conflict and other cultural, religious, social and political tensions with a series of tranquil experiences.[43] In this fantasy land, rock exists only to be climbed, not quarried—never mind that much of the city is constructed with quarried material.

The problem with this vision of Delhi is that it does not actually solve the problems plaguing the city; it just moves them off to a less conspicuous location. The city's population continues to rise at an unsustainable pace, but walls and cordons, both figurative and literal, keep green areas green, spacious areas spacious, poor areas poor, polluted areas polluted. Mining continues, but on the sly, or simply further away. The beneficiaries of these spatial re-arrangements now enjoy a "vibrant" city of spotless parks, gleaming malls, mountain biking routes and rock-climbing outings. But something is missing. An amusement park ride may be fun; but to live one's days in an unending amusement park ride is, finally, nauseating. And there is a sense of unrest, a feeling that the walls and cordons will eventually break under so much pressure.

Thankfully, this vision of the city is relatively recent and, if geology has taught us anything, it's that everything changes, even those things that seem most permanent. Perhaps the geology metaphor can be pushed even further. Geologists speak of the Aravalli range "suffering deformation" in the process of tectonic rifting and shifting. But this process eventually led to a grand mountain chain, one composed largely of metamorphic rock, something entirely new, but with traces of the old shot through it. And even the decay and weathering of the Ridge has led to surprising opportunities. The city of Delhi has certainly seen its share of suffering and deterioration, along with a series of metamorphoses. If the Disneyland version of Delhi is a chimera, can something more promising rise in its place?

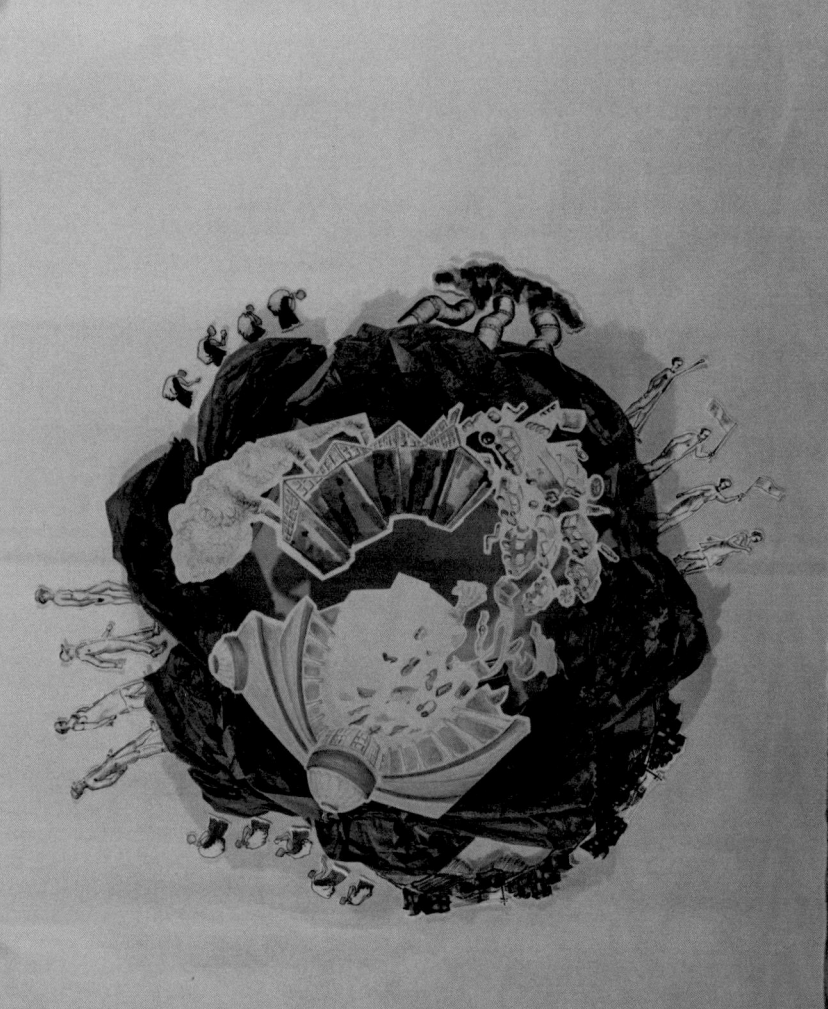

The Ridge as site of
Production & Consumption

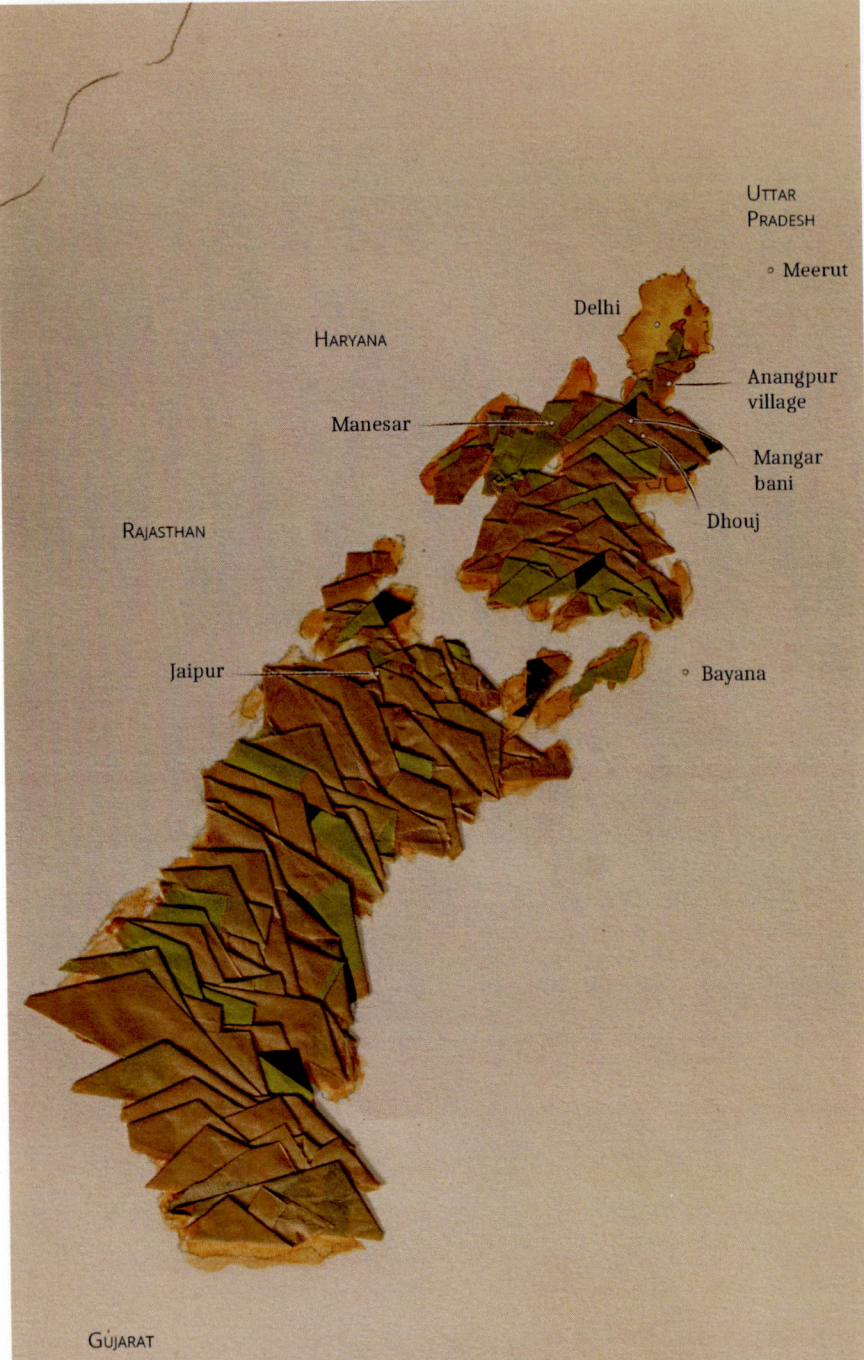

The Aravalli Mountain Range

Signature Bridge

Northern Ridge

Shahjahanabad

Central Ridge

South Central Ridge, Sanjay Van

Ashokan edict

Vasant Kunj malls

Gurgaon

Israil Camp

Tughlaqabad

Lado Sarai

Mehrauli

Southern Ridge

The Aravallis enter Delhi

Ridge-as-Utopia

Even given this hybridity, though, communities with pastoral roots have had an outsize impact on Delhi. This is largely because pastoralists are, on the whole, a more mobile bunch. Pastoralists entering the Indian subcontinent from arid Central Asia have played a crucial role in the shaping the Ridge and the city of Delhi. The history of these transformations, which this chapter seeks to tell, is one in which ecological and social history are intertwined. It's about the Ridge's stony soil and what has emerged from that soil; it's about those migrants who put down roots in Delhi and begin to consider themselves "sons of the soil."

If the previous chapter was an eclectic geological history of the Ridge, here is an eclectic ecological history, with human communities fully integrated into the story of the land. Although pastoral groups are central to this drama, they are just one part of a larger cast of players, which includes saints, bandits, warriors, merchants, pilgrims, farmers, landlords, administrators, miners, politicians, lawyers, activists and many more. All these figures have sprung up on the soil of the Ridge.

The Rise (and Fall) of the Gujjars

In this book's introduction, we witnessed the mythical burning of the Kandava forest by Aryan pastoralists, which hints at the long history of pastoral livelihood strategies in the Delhi region. There is a broad scholarly consensus that Aryans migrated into northern India from central Asia, bringing with them their pastoralist lifestyle and their pastoralist gods (including Agni, the god of fire).[2] This consensus has been fiercely contested by Hindu nationalists, who insist, against the weight of mounting evidence, that the Aryans must have originated in India; but the controversy around the so-called "Aryan migration theory" should not distract us from the fact that Aryans were just one of many groups passing through the subcontinent's northwest corridor, entering India from Central Asia.

From the perspective of the Delhi Ridge, perhaps the most important migration was the one that brought the Gujjar community to India. More than any other group, Gujjars have laid claim to the title of "sons of the soil" on the Delhi Ridge, and they still play a vital role in shaping the social and ecological contours of Ridge land.

Initially, though, Gujjars were just one more stream in the unending flow of peoples and cultures entering India through the northwestern corridor. The evidence here is a bit shaky; most of it is from the colonial era and it relies on rather dubious etymology. There is speculation that Gujjars originally came from Georgia (re-written as Gurjia or Gurjaristan), or that they broke off from the Gujj tribe of southern Khurasan in Central Asia. Despite these confusions, there is a fairly solid scholarly consensus that the Gujjars were a pastoral, nomadic tribe that moved from Central Asia to South Asia sometime in the early or mid-first millennium AD, establishing a set of roving communities mainly in present-day Rajasthan and Gujarat (in fact, this is where Gujarat gets its name).[3]

The picture gets a bit clearer once we reach the seventh century. By this time, there is evidence that some segments of the Gujjar community had given up the nomadic pastoral life in favor of full-time agriculture. Meanwhile, other Gujjar groups were drawn into the powerful currents that were reshaping the political landscape. Most dramatic was the fate of the Pratiharas (as known as the Gurjara-Pratiharas), a Gujjar clan that considered itself Rajput. The rise of the Pratiharas parallels the larger ascension of Rajput clans as dominant political and social players in Rajasthan (formerly known as Rajputana) and eventually in the Delhi region.

Although, in the present day, the term "Rajput" conjures up the image of a fixed, high-status, closed caste group, the emergence of the Rajputs in early medieval Rajasthan suggests a more fluid, ambiguous category. "Rajput" was a broad category into which various individuals and tribes could fit as they transitioned from a relatively isolated tribal life to a position at the head of emerging states. It may be more accurate, then, to speak of a process of Rajputization, rather than a stable social entity called "Rajput".[4]

Further, the emergence of the Rajput was inextricably tied to changing livelihood practices, and particularly the expansion of the agricultural economy. Put simply, states favor sedentary agriculture over nomadic practices (whether pastoralism or hunting and gathering). The apparatus of the state—capital cities, armies, an expanding number of rulers and their retinue—needed the surplus food and goods that settled agriculture could provide. By bringing more land under agriculture, states could increase the surplus at their disposal and build their military, economic and social strength.

Such were the dynamics at work as Rajput groups began to form. Initially, the term "Rajput" mostly referred, not to kingly lineages, but to individuals who joined established armies as mercenary soldiers, and then used their military status to become landowners and if possible establish their own power center. Eventually, "Rajputs" began to emerge as distinct social groups, first as minor powers supporting existing kingdoms, then as independent powers developing their own networks of alliances, marriages and dynasties. The array of groups given the designation "Rajput" in medieval times (by no means a fixed list) included tribes as varied as the Medas, a group that had long resided in the area, and the Huns, a group of clearly foreign extraction that had recently assimilated into the local communities.[5]

The Gurjara-Pratiharas were one of the most powerful Rajput clans, and they emerged from their local context to play a decisive role in North Indian politics. Early references to the Pratiharas suggest that they took part in the expansion of settled agricultural kingdoms at the expense of smaller, more mobile tribal settlements. This was often a violent process. A Pratihara leader known as Kukkaka is credited with raiding the cattle of other tribes, burning down the enemies' villages (echoes of Arjuna and Krishna) and establishing markets for merchants.[6] By eighth century, the Pratihara empire had expanded significantly, and had moved its base of power from present-day Rajasthan to present-day Uttar Pradesh, and specifically the city of Kannauj, which has a long imperial legacy.[7]

From Kannauj, the Gurjara-Pratiharas made alliances and expanded their power throughout North India, including Delhi. A Rajput clan called the Tomars controlled the Delhi region, likely first as generals of the Gurjara-Pratiharas, then as governors ruling in the name of the Pratiharas, and finally as independent kings.[8] According to Gujjar tradition, the Tomars were also of Gujjar descent, perhaps cousins of the Pratiharas, and their line continues to this very day with the Tanwar subcaste, which is prevalent in many areas of the Delhi Ridge.[9] (These claims of ancient lineage are often made in strikingly modern ways. An online platform for the Gujjar community contains threads with titles like "Founders of Delhi, the Gujjars Rise Again".)

Whatever their provenance, the Tomars were to inaugurate the tradition of building walled settlements in Delhi, using quartzite mined from the Ridge to create structures on the heights of the Ridge. The Tomars constructed their fortress, called Lal Kot, in the southwestern corner of the Ridge, most likely in the mid-eleventh century. Lal Kot is now in ruins, its crumbling ramparts dotting Sanjay Van park.

But one hardly needs to wait a millennium to appreciate the transitoriness of the Tomar's reign. Soon after emerging as an independent power, the Tomars suffered defeat at the hands of yet another Rajput power: the Chauhans. Little is known about Vigraharaja IV, the Chauhan king who toppled the Tomars. But his nephew, Prithviraj Chauhan, also called Rai Pithora, has become a key figure in Indian history and mythology. Under Prithviraj's rule, the fortifications of Lal Kot were extended, again using the quartzite blocks of the Ridge, to create an enlarged fortress known as Qila Rai Pithora. However, Prithviraj did not use this as his main base; Delhi was a military outpost manned by one of his vassals.

Despite this fact, Prithviraj is lauded by Hindutva historians as the last Hindu king of Delhi, renowned for his clashes with Mohammad Ghori, sultan of the Central Asian Ghurid empire. Prithviraj defeated Ghori's army in 1191, but fell to the same army in the following year. 1192 is thus given great importance by Hindu nationalists. It is regarded as the year of the fall; this is when the "outsiders" finally succeeded in breaching the defenses of the pure Hindu subcontinent.

Ferment in the Arid Zone

This may be questionable history, but there's no doubt that Ghori's victory, and the increasing importance of Delhi in this period, point to significant shifts. The emergence of Delhi's Ridge-top fortresses, and the rapidity with which they changed hands, are indications of a wider churning. In this period, Delhi became a crucial hub in the intersecting lines of trade, transport and warfare that were spreading throughout a huge part of the world.

These economic and political changes cannot be separated from Delhi's ecology, and its location within a much larger ecological zone. Delhi stands between the extreme dryness of Rajasthan to the west and the well-watered Gangetic heartland to the east. To its west lies not just the dessication of the Thar desert, but also the sprawling, discontinuous ecosystem that the historian Jos Gommans refers to simply as the Arid Zone.[10] This area includes all the dry zones stretching from northeastern Africa across the Middle East to Central Asia, up to Southeastern Europe and down to South Asia; roughly, it comprises all the areas in these regions that receive less than 1,000 millimeters of rainfall annually. The Indian subcontinent as a whole serves as a transitional zone between this drought-prone area and the lushness of monsoon Asia. Delhi and its Ridge play this role in miniature.

Even within the Arid Zone, of course, there were oases of agriculture, largely along rivers. However, in most of this zone, nomadic pastoralism had the upper hand. With such little rainfall and such unforgiving soil, a nomadic lifestyle has distinct advantages; pastoralists could easily move between different grazing areas depending on the vagaries of rain patterns, and their livestock could survive on

The Arid Zone

the grasses, shrubs and stunted vegetation that proliferated throughout the zone.

The ninth and tenth century saw an expansion and a subsequent melding together of many different nomadic groups in the Arid Zone, creating a rich cultural mix that drew on Arabic, Persian and Turkic traditions. This culture, and the technologies that went along with it, then spread into India, with mastery of war-horses an especially important element. Though horses had existed in the subcontinent since the time of the Aryans, new migrants from the Arid Zone brought both new breeds and new styles of mobile warfare, in what Gommans dubs a "horse-warrior revolution".[11] More broadly, this was an age of pastoralists, merchants and pilgrims criss-crossing the Arid Zone and creating disruptive new flows of ideas, technologies and armies.

Although many of the horse-mounted warriors entering the subcontinent were Muslim, what impressed local populations was less the warriors' religion and more their equine accomplishments. Rajput rulers were among the first to adapt this new military technology, which greatly aided their expansion. Two other types of animals were also important: camels and cattle. The latter was important both for long-distance transportation (of people and of goods) and for agriculture (plowing the fields). The importance of cattle gave increased prominence to pastoral communities, who had extensive experience breeding hearty cattle and guiding them over long distances. On the backs of horses, camels, and cattle, rising powers from the peripheries of the Arid Zone were able to establish new routes for trade, pilgrimages and plunder.

The kingdoms emerging in this era had to maintain a fine balance between the newfound resources of the Arid Zone and the riches of the agriculturally fertile zones whose powers they were displacing. They relied on nomadic groups to work as cattle-breeders, as merchants, and as mercenary soldiers; but they also needed agricultural surplus, growing cities, and densely settled populations to tax and control. This explains the location of many cities that arose in this time period, cities that sat on the border between dry pastoral zones and more fertile agricultural ones. In northwestern India, Delhi is the prime example of such a city. The Ridge was an especially attractive location for new kingdoms since its imposing heights had clear military advantages.

This big-picture perspective helps us recast the lines of conflict that have plagued both India as a whole and the Delhi Ridge in particular. This was not, as the Hindutva narrative avows, a "native Hindu" vs. "invader Muslim" conflict; the larger tension was between settled agricultural powers with their increasingly parochial Brahminism and newer powers that drew on the resources of the Arid Zone, established cities in transitional zones, and championed emerging religious movements that ranged from Sufism to the cult of Rama. True, many of the newer powers embraced Islam. But many didn't, including the Rajputs, who also took advantage of the resurgent Arid Zone, and who helped establish intersecting trade and pilgrimage routes throughout the subcontinent. The post-1192 rulers of Delhi, broadly referred to as the Sultanate, should be understood in this context.

Rulers vs. Gujjars

The Delhi Sultans, who themselves had emerged from pastoralist backgrounds, had an ambivalent relationship with the pastoral communities that surrounded the city, particularly concentrated in the arid zones to the west of the city's new walled fortresses. Delhi's new rulers and the pastoral communities fell into an uneasy alliance, one which was marked by cooperation and collaboration, but also by conflict and distrust.

The settled state often tried to expand into pastoral zones and to convert grazing lands to agricultural fields; the introduction of canals was a crucial strategy for creating new zones of fertile soil. Ideologically, too, the state often downplayed the role of pastoral groups, preferring to laud the agriculturalists from whom surplus could be extracted more easily and who were generally more pliable. Pastoral groups resisted state control, and given the frequent attempts to undermine their livelihood, they felt fully justified in conducting the occasional cattle raid or act of banditry.

This dynamic is especially clear with the Gujjar community, one of the dominant pastoral groups in the Delhi area. In recent years, some scholars from the Gujjar community have adopted the Hindutva perspective to explain the Gujjars' decline in medieval India. For these scholars, 1192 was not just a disaster for India, it was a specific tragedy for the Gujjar community. Gujjar history has now been retrofitted, made into a glorious (if doomed) struggle against vicious outsiders; a recent book about the community is entitled *Heroic Hindu Resistance to Muslim Invaders (636 AD to 1206 AD)*.[12]

This view of history draws on the imperial grandeur of the Gurjara-Pratiharas, and on the Tomars who ruled Delhi in their name. The victory of Mohammad Ghori over Prithviraj Chauhan is recast as a victory of Muslims over Gujjars. However, by the late twelfth century, the grand Pratihara empire, which once stretched across most of northern India, had already disintegrated, leaving in its wake a fractious set of small warring states. Given the history of the resurgent Arid Zone, the downfall of the Pratiharas may very well be linked to their abandonment of pastoral ways and pastoral zones, and their establishment of an imperial capital well into the humid, fertile zone of the Gangetic Basin. They could not withstand the rise of newer powers, including both their rival Rajputs and Central Asian forces.

Even more importantly, the royal, Rajputized Gujjars were only ever a tiny fraction of a larger Gujjar population that had spread throughout the northern half of the subcontinent. This population was internally stratified to an enormous degree, with many communities continuing the age-old tradition of nomadic pastoralism, some making the transition to settled agriculture, and a select few grabbing power across the region.

The extreme heterogeneity of Gujjar groups makes it difficult to establish a connection between the royal Tomars of the eleventh century and the present-day Tanwars in Delhi. For the vast majority of Gujjars of northern India, both before and after the coming of the Muslims, their lives were dominated by either pastoral or agricultural pursuits, or, in many cases, a mix of both. A new set of rulers did little to change that.

Still, it is true that the overall reputation of Gujjars suffered in the Sultanate period. After the fall from imperial grace, they were certainly perceived as a threat to "settled" civilization. A celebrated religious tale from this era shows the casual disdain heaped upon Gujjars. The story is set in the fourteenth century, when the emperor Ghiyasuddin Tughlaq was building the new capital of Tughlaqabad

agriculturally productive, it could not be taxed, and it was, in British eyes, mere detritus. But this pejorative term obscures the range of ecological functions "wastelands" perform. For many groups, and most importantly for pastoralists, these "wastelands" were essential. They included both wooded areas and open grasslands and served as important grazing routes for various pastoral communities.

Like previous empires, the British Raj had many reasons to promote agriculture and many reasons to fear and suspect nomadic pastoral communities. In their singular drive for profit, however, they promoted agriculture more aggressively and more extensively then their predecessors, especially as their grip on the region strengthened. The British were further emboldened by their unwavering belief in the "magic of private property", a phrase much used at the time. In their own country, they had seen how the death of the commons and the rise of the individual, propertied farmer had led to huge increases in agricultural profits. (It mattered little to the British elite that this burst of efficiency happened at the expense of the great masses of agricultural workers, who were thrown off the land and formed the bulk of the growing urban proletariat.) It was the British goal to unleash just such an agricultural revolution in India, for their own benefit, of course.

But their short-sightedness, and their overriding emphasis on profit, led not to agricultural revolution, but to stagnation and decline. The earliest efforts were in Bengal, where the East India Company implemented its first "Permanent Settlement" in 1793. Such settlements were a key instrument in the colonial toolbox. They were based on detailed assessments of the land, and they determined the amount that the natives would have to pay in agricultural taxes. The hope was that, by clearly demarcating which land belonged to which landlord, and by keeping a fixed assessment rate, the land settlements would lead to a flurry of agricultural improvements. But the settlements usually had just the opposite effect. East India Company officials established taxes at an unrealistically high level. Farmers were not able to pay. They got trapped in a cycle of debt, agricultural productivity fell, and rural discontent grew.[21]

By the time the British began their Land Settlement efforts in Delhi, they at least recognized that their previous attempts had been misguided. In Bengal, and then in Madras and other parts of British India, administrators had first targeted individual landlords, and then individual peasants, as the potential payers of land revenue. In Delhi and the surrounding regions, they struck upon a new formula, drawing on the results of the newly-emerging field of anthropology. India, they posited, was a land of village republics. And indeed, in the Delhi area, they found a strong tradition of village councils composed of landowning males. If these councils already existed, why not just tax them?[22]

Such was the thinking when officials began Land Settlement efforts in Delhi District in 1817. But the British could not solve the riddle of Indian society (and how to make it pay) so easily. Since their focus was on village communities as isolated entities from which revenue could be extracted, they ignored the connections between various villages, and between settled villages and mobile pastoral groups like the Gujjars.

Their blindness resulted not just from avarice, but from fundamental assumptions about land. Despite their ostensible recognition of village councils and community ownership of land, the British could not escape a worldview dominated by private property. For them, land was a commodity, something to be bought and sold. They introduced rules that determined the landholdings of individuals based on their current cultivation (as opposed to the much older custom of ancestral plots), and they introduced methods for privatizing commonly held lands. Most importantly, they insisted on drawing sharp boundaries between villages where before there had been much fuzzier borders. Especially in pastoral areas, there had been little effort to demarcate the "waste" between villages and within villages. The haziness of precolonial frontiers was an endless headache for the British, since it made the formal sale of land remarkably complex.

The simple act of drawing exact boundary lines had profound effects on the hinterlands of Delhi. These lines tore the connective tissue joining pastoral and agricultural collectives and erased the customary rights of nomadic groups. In the early days of British rule in Delhi District, the main emphasis was expanding agriculture at the expense of pastoral activity. So, once boundary lines were drawn, agriculture was promoted in "waste" areas that fell outside the new village borders, often in zones frequented by pastoralists. This process was sped along by the construction of canals throughout the district.

The government also tilted the scales against pastoralists by encouraging intensive agriculture within villages, and setting up tax rules that punished the practice of keeping land fallow for long periods of time. Such "long fallows" were key sites both for pastoral activity and for the regeneration of the soil, but in their zeal for agricultural production, the British were happy to overlook these factors.

The British offensive against pastoral livelihoods was not simply a matter of tax policy and technological changes like canal construction. British officials also waged an ideological war against the whole idea of nomadism in general, and pastoralism in particular. Perhaps this is no great surprise. Rulers in India from the Sultans onwards were to curse the rebellious, independent, and unpredictable nomadic groups that refused to be good, tax-paying citizens. An official British description of a pastoral zone in Delhi District sums up this contemptuous attitude nicely: the area in question is "overgrown with grass and bushes, scantily threaded with sheep walks and the footprints of cattle" and the "chief tenants are nomad pastoral tribes who knowing neither law nor property collect herds of cattle stolen from the agricultural districts."[23] Any rural group that adopted non-agricultural means of livelihood was tarred as hopelessly lazy and criminally inclined.

This rhetoric took on a particularly nasty tinge as imperial theories of race began to gain ground. Not long after Darwin published his revolutionary theories of evolution, a crude, politically motivated application of Darwinism sprang up. Known as Social Darwinism, this ideology applied the concept of evolution to different human communities, venturing that different groups were engaged in a "survival of the fittest", and that some had a distinct advantage. Drawing on dubious science, the groups were theorized as races that were biologically distinct, sharply differentiated by the process of evolution. While these supposedly biological differences were most easily seen in physical markers, such as skin color and skull size (hence the obsession with phrenology), the Social Darwinists linked these physical differences with intellectual, emotional, and civilizational differences, creating a hierarchy with the European races (of course) at the top.

In India, racial language was used to identify particular groups that were not sufficiently pliable and had resisted British rule. The apotheosis of this strategy was the Criminal Tribes Act passed by the British Indian government in 1871, in response to an uptick in banditry in the countryside. Suffused in racial language, the Act identifies certain "tribes, gangs and classes" that are "addicted to the systematic commission of non-bailable offenses". It specifies that such groups must be prevented from practicing "their hereditary professions of theft, robbery and dacoitry". Simply being born into one of these groups had become a criminal act.

The Gujjars had a prominent place on the list of Criminal Tribes, and the attitude of the British toward Gujjars was unfailingly derogatory and often couched in racial terms. The Delhi District Gazetteer, published by the British government in 1884, stresses that thievery and deception are ingrained in the character of the Gujjar. Although noting that they are relatively light-skinned, the Gazetteer goes on to remark that "the Gujar has generally been a mean, sneaking, cowardly fellow, and it does not appear that he improves much with the march of civilization." But beneath the racial slurs, there was always a level of economic calculation. The Gujjars are derided as "third-rate cultivators" who contribute nothing to the British drive to expand agriculture.[24]

Entirely absent in colonial accounts of the Gujjar community is any recognition of the role the British themselves played in antagonizing Gujjars and pushing them towards extra-legal activities. Successive land settlements and redrawn village boundaries made it increasingly difficult for Gujjars to make a living as pastoralists. While cattle raids had long been a feature of pastoral life, the nineteenth century saw a huge increase in the number and organization of raiding groups; despite the racial language used by the British, these groups were often quite diverse, with members drawn from different castes and religions, united in their economic distress and their resentment of British rule.[25] In their minds, they were not petty thieves, but rather noble bandits.

In part due to government policies and persecution, many Gujjar groups in the Delhi area had become settled by the mid-1800s, practicing agriculture alongside more traditional pursuits of cattle-rearing. But even this was not enough to guarantee a secure life. On several occasions, their land was simply snatched from them by the government and its officials. Representative in this regard is Thomas Metcalfe, who served as Resident of Delhi, and who took land from a series of Gujjar-dominated villages stretching from Shahjahanabad to the Northern Ridge. One of these villages was Chandrawal, which, at least in Rahul Khari's account, had already faced similar pressure at the hands of the Mughals. Metcalfe used his new property as the site of "Metcalfe House", the most expensive mansion in northwest India at the time. On the southern part of the Ridge, Metcalfe took over the tomb of a sixteenth-century Mughal nobleman and used it as his personal retreat, as well as renting it out to honeymooning couples. These land grabs only added to the perception of the British as an alien power intent on undermining the livelihoods and lifestyles of local populations.

These frustrations came to a head with the Great Uprising of 1857. Historians continue to debate the precise causes and contours of the uprising, but there is widespread agreement that rural discontent, often stemming from British agrarian policies, played a major role in turning the conflict from a military mutiny to a widespread rebellion against British rule.[26] This is certainly true in the Delhi region. One of the most striking acts of the rebellion was the burning of Metcalfe House by the villagers of Chandrawal. After the British brutally crushed the uprising, the Gujjars paid dearly for their role in it. All of Chandrawal was confiscated by the government; it was briefly rented out to non-Gujjars as grazing land and was eventually sold off in plots intended for British bungalows. The surviving villagers moved to a site further north, bringing the name of the village with them; they resumed their pastoral livelihoods as well as they could and tried to rebuild a sense of community life.

From "Waste" to "Forest"

The vengeance against Gujjars and other rebel communities in 1857 was swift, brutal and undeniable. Its wounds were immediately apparent to those on both sides of the conflict. The slow draining of the life-force of the land, which both preceded and followed the Mutiny, was more difficult to discern. The land was transformed in both dramatic and mundane ways. Canals proliferated; new villages were created in former "wastelands"; long fallows were converted into short ones or into constantly-tended fields; land was redefined as a commodity and a market for land thus established; and common lands were privatized and sold to the highest bidder. These changes inevitably had wide-reaching effects: nomadic tribes were forced to settle; tensions between agriculturalists and pastoralists flared up with increasing regularity; farmers unable to pay extortionist tax rates fell into poverty and debt; ancestral properties were transferred from the old landed class to a new set of middlemen, mainly bankers and urban professionals.

These profound changes were felt most acutely in pastoral zones, as the entire pastoral way of life was threatened. At times, the British were forcefully confronted with the damage they had wrought, most vividly with the rise of famines. In the nineteenth century in Delhi District, droughts, scarcities and famines became alarmingly common in pastoral areas, and the government eventually began to recognize the collateral damage their policies were causing. By the end of the century, the drive to extend agriculture had slowed considerably.

But the Raj was not simply motivated by humanitarian concern or by fear that a famine-prone region might be inclined to start another 1857-type rebellion. It was increasingly recognizing that the "waste" in British India was, perhaps, not so useless after all. This belated realization was largely driven by technological change. The railroad, that potent symbol of modernity, was introduced in the region in 1862. As the rail system expanded, it needed to be fed with the forests of India. The construction of railway lines required a huge amount of wood, mostly for the sleepers that supported the tracks. For this wood, administrators largely turned to the "wastelands", in which local populations had maintained many forested zones.

Teak and sal, the trees used for sleepers, were not found in Delhi, although they were available in the larger Punjab Province, of which Delhi District was a part. But closer to the Ridge, "wastes" provided essential ingredients for building streets, which complemented the rail network as British influence penetrated deeper into the subcontinent. These ingredients were mineral: the stones, pebbles and sands (including Badarpur sand) used by the British Public Works Department for road construction. As early as 1824, the government introduced laws allowing them to temporarily acquire "wastelands" rich in mineral wealth, and to dig up whatever they needed before returning the land to the locals.

By 1878, it was clear to the British that they needed a stronger administrative tool for controlling the "wastelands". The Indian Forest Act, introduced in this year and revised significantly in 1927, was this tool. It allowed the government to set aside any patch of land as a "Reserved Forest" and to prohibit the cutting of trees, collecting of firewood, grazing of livestock, and other activities. Of course, the government itself was free to cut down trees in protected forests, as it did with great zeal to build its empire. It could also selectively allow some grazing and other pastoral activities, after charging a hefty fee.

In one sense, the Indian Forest Act can be seen as a reversal on the part of the government, a recognition, at long last, of the value of so-called "wastelands". But in another, more fundamental sense, the Act was merely a continuation of the state's longstanding policy of introducing radical changes in land ownership practices with an eye to filling government coffers. The Indian Forest Act was not really about ecological concern, but rather about control over land, and thus control over valuable resources.

The government's instrumental view of ecosystems—evaluating the land based on the presence or absence of economically valuable trees—led to a fundamental misrecognition of the Indian landscape. This, at least, is the implication of new ecological research that argues for a wholesale recategorization of Indian forest types, which even today follow British typologies. This research, by the ecologist Jayashree Ratnam and others, calls into question British descriptions of wide swaths of India, including Ridge habitats, as "dry deciduous forests". The British emphasis on trees, as opposed to entire ecosystems, blinded them to the fact that "dry deciduous forests" were, in fact, savannas, that is, mixed tree-grass ecosystems in which the grass species actually play a driving role.[27] The British categorization is so deeply ingrained that it is still used by the vast majority of environmentalists in India, and it has only recently been questioned by a small group of scientists.

But the British disregard for the environment went well beyond misclassification. Even though the value of trees was slowly dawning on them, this recognition could still be overruled by more pressing economic concerns. The actions of the government after the implementation of the Indian Forest Act hardly suggest environmental enlightenment. From 1900 to 1920, 3,000 square miles (7,770 square kilometers) of forests were cleared for agriculture in Punjab Province, an area that then included Delhi.[28] To put this into perspective, this is one hundred times the area currently designated as Reserved Forest in the Delhi Ridge.

A Tale of Two Settlement Officers

Ironically, in this same period of time, two areas of the Delhi Ridge, though largely denuded, were classified as Reserved Forests. This designation built on twenty-odd years of largely unsuccessful efforts by various colonial administrators to afforest the Ridge, which had been suffering from both the long history of urbanization in Delhi and the recent introduction of detrimental land settlement policies. Even more ironically, the government officials who carried out the ecologically disastrous land settlements were the very ones who proclaimed the value of "forests" most loudly.

Here we're not talking about a big cast of characters. In fact, we're just talking about two officials. One of the most striking aspects of British rule in India was the staggeringly large set of responsibilities given to single individuals, a reflection of both the drive to cut administrative costs and the hubristic presumption that one officer could easily navigate the innumerable complexities of Indian economic, political and social systems.

So when it came time to conduct land settlements in Delhi District, just one officer was entrusted with the task, along with a small team of assistants. Although the very first land settlement was in 1817, the most comprehensive ones were much later, in 1880 and 1910. The 1880 settlement was conducted by J. R. Maconachie, an official who had worked on land issues in Delhi District and the larger Punjab Province for over two decades. In the records that he left behind, Maconachie comes across as an enthusiastic, even kind-hearted administrator, but one nonetheless biased by the prejudices of his time. He clearly had a deep knowledge of the region, publishing one book called *Selected Agricultural Proverbs of the Punjab* and another on customary law in Delhi. He was also keenly interested in forestry, and one of his chief missions as an officer was to increase the tree cover in Delhi District and beyond.

And yet, in 1880, he authored a land settlement report that continued the trend of prioritizing agriculture, demonizing pastoralism, and weakening collective institutions that had previously played such a key role in maintaining sacred groves and other forest patches. What is more, he did so with a clear conscience, with the (mis)understanding that such measures would actually have a positive impact on forest conservation.

Maconachie's main mistake was his inability to locate the true roots of the environmental crisis, which made him confuse cause and effect. This was not due to thoughtlessness or carelessness; Maconachie was conscientious and thoroughgoing in all his reports. Rather, Maconachie's vision was narrowed by the ideological blinders he wore. He viewed pastoralists with great suspicion and held them to be the main cause of the deforestation rampant in Delhi and the larger region.

His settlement report reflects his biases. The report was no mere cataloging of traditional land tenures and soil productivity assessments; it was also an opportunity for Maconachie to opine on a wide variety of social and racial matters. He joined his fellow officials in his outright disdain for Gujjars. He notes that in Delhi District, the Gujjar community "has appropriated almost entirely the hill villages, as they suit their pastoral traditions, and pastoral traditions are less repugnant than a settled husbandry to thieving, a habit universally attributed to the Gujjar."[29] He prefers the Jats, traditional rivals to the Gujjars, who were more deeply rooted in agricultural traditions.[30] He waxes idyllic, saying "Nothing is pleasanter, of its kind, than to walk through a well-cultivated Jat village."[31]

But he has not given up all hope for the Gujjars. He maintains, "There seems reason to hope that a material improvement in the habits of the Gujjar is setting in. The agriculture of the hills will be greatly aided by the *bands* now being made or repaired; and this will probably in itself prove an inducement to pursue the path of honesty."[32] *Bands* or "bunds" are earthen dams used to collect water and provide sources for irrigation; they were widespread in pre-colonial times, and Maconachie shows unusual sensitivity in his appreciation of these traditional systems of water management. And yet his statement on the whole reeks of standard-issue colonial ideology: Gujjars can only be "induced" to be honest if the Ridge can be converted from a pastoral to an agricultural zone.

Shortly after writing this report, Maconachie was able to test out his theory on the northern portion of the Delhi Ridge. His goal was to stabilize the soil using bunds. In his mind, this would have two distinct positive effects: an increase in the number of trees that could be planted, and an extension of the agricultural zones of Delhi into the hills. For his test site, he chose a part of the Ridge close to the village of Chandrawal, which was, one may imagine, not in the mood for any more state intervention. Not that the villagers had a say in the matter.

Maconachie was perhaps the most dangerous kind of colonial official: one who genuinely believed he was doing good for the locals. He was not, as far as the records show, animated by the naked greed that motivated many early colonial adventurers. He instead played the role of the stern but loving father, who always knew what was best for his children.

In his bund project, Maconachie tried to recruit the men of Chandrawal to help with his construction efforts. It is unclear how successful he was, but, given that the project was abandoned within three years, it is unlikely that he found support for a key component of his plan: forbidding grazing in the areas to be afforested. A letter on the plan states,

> The first measure is to make arrangements for effectually excluding goats and cattle from browsing on it... The villagers who own any plots which may be needed would have to be dealt with, but they... I trust will be induced to consent to what will ultimately benefit them, as covered with wood, the ground will be more valuable than it is now.[33]

Maconachie was not helped by larger trends tied to the regrowth of the city after the devastation of 1857. From 1880 onwards, there was a sharp increase in the demand for meat in Delhi; it is understandable that the Gujjar villagers of Chandrawal preferred the guaranteed income of a consistently growing meat market, rather than the vague promises of government officials for "valuable" forest land, especially given both the overt and covert violence of the state against them.

Urban expansion was thus the backdrop to the 1910 settlement report, which was carried out by none other than Major Henry C. Beadon, the protagonist of Chapter 1's quarry quarrel. Beadon, like Maconachie, was deeply interested in forestry, and, like Maconachie, he was convinced that the main enemy of the forest was livestock, and more specifically, the goat. In his settlement report, he notes, "The great increase in sheep and goat is remarkable and from the point of view of new arboriculture deplorable, as the main increase is in goats: flocks are a profitable income to the owners who graze them unscrupulously where they choose."[34] By "new arboriculture", Beadon simply means the planting of new trees. Like so many before him, Beadon creates a simplistic opposition between not-to-be-trusted villagers and wise British administrators who knew the value of trees.

In an earlier memorandum, Beadon had taken the opportunity to rail against goats at greater length. Discussing the hills at the southern end of Delhi District, Beadon maintains,

The main destructive agency there, and I am convinced everywhere else, is the goat. It is most unfortunate that goat-keeping is lucrative: but there is no getting over it, and the result is that wherever there is waste and uncultivatable land, that area is overrun with goats who browse down every tree and every shrub.[35]

He goes on to stress the importance of planting tree saplings in the hills but cannot resist take one more swipe at goats: "The greatest remedy of all in my opinion, however, is the strict control of goat grazing. There are people who hold that the goat is a wonderfully economical animal: so it is, as it lives at the expense of posterity."[36]

Beadon, of course, is on the side of posterity. Even more than Maconachie, he is animated by an unshakable conviction that he is part of a civilizing mission, making tough choices and noble decisions on behalf of the native population. After his repeated anti-goat harangues, he concludes with this rousing, patriotic sentiment:

All forest conservation is locally unpopular, but a time arrives when the parting of the ways is reached and the State has to decide whether it will follow in spite of difficulties a firm policy of conservation, or the craven policy of inaction or half measures which are little better than inaction. If France in 1860 and 1864, during the disturbed era of the Second Empire, could take a strong line with success, there can be no excuse for a weak policy in British India.[37]

For Beadon, then, prohibiting grazing, and stigmatizing pastoralists in the process, was nothing less than a sacred civic duty.

In a way, Beadon and Maconachie were right; grazing did prevent trees from springing up on the land. The tension between grazing and forestry had, after all, been noted since the time of the *Mahabharata*. And there is ample evidence that, in areas like the Ridge, if grazing and other human activities stop and the land is left to itself, forests (or, more accurately, tree-filled savannas) will eventually emerge. In recent decades, this has been seen again and again in the former mines along the Ridge, which were themselves former grazing grounds. If these closed mines are left alone, whether through neglect or through active conservation efforts, diverse ecosystems grow back with surprising speed.

But the British officials were viewing the goats vs. trees tension through far too narrow a lens, ignoring the tumultuous changes to the land that they themselves had inaugurated. If given the opportunity and the land, pastoralists were very well capable of maintaining thriving forests along with grassy grazing lands. But the British had denied them that opportunity, or rather had taken it from them piece by piece. British land policies, with their exorbitant tax rates, firmly defined boundaries, and obsession with enlarging agricultural zones, were an unprecedented affront to pastoral life and to the groves carefully preserved by pastoralists.

By 1880, when Maconachie turned his attention to the Ridge near Chandrawal, the zone had long been deforested. Whether this was due to the urban pressures of Shahjahanabad or the policies of the British is difficult to know. But for colonial officials, the solution was clear. What was needed was more state control, not less; more expropriation of land from locals, not less. They were intent on punishing pastoralists for conditions beyond their control.

Maconachie's actions in this regard were relatively sensitive; paternalism aside, he at least tried to get the support of the local community and involve them in the tree planting process. After Maconachie, a more brusque approach was favored. Leaving aside any pretense of community uplift, later administrators stressed that the area should be afforested for the benefit of nearby British residents, with local communities strictly excluded. Officials began to construct fences around these areas. They also started conducting cattle raids, unaware of the irony that they were practicing the very activity that supposedly proved the 'savagery' of the Gujjars. One official proudly reports,

I have carried out, by night and day, various raids on the trespassing cattle, and by not releasing the cattle on the payment of the merely nominal fine but prosecuting the owner for mischief I have, I hope, partly succeeded in convincing the latter that for the present at any rate it will pay them better to graze their cattle elsewhere.[38]

But these attempts were not as successful as the administrators had hoped. The Gujjar villagers would not abandon their means of livelihood so easily. They found many ways to subvert the new system of exclusion, to the exasperation of British officialdom. A typical complaint in 1909 reads, "The grazing can be stopped, but the difficulty will be in preventing grazing at night, as these Chandrawal men are not above turning their cattle in to the public gardens and private compounds at night."[39] But despite intense local resistance, the British kept up their efforts, building fences, hiring guards, conducting cattle raids and introducing regulation after regulation to prevent grazing.

In 1915, the transformation of the Northern Ridge was finally complete; under the Indian Forest Act, the area Maconachie had originally tried to afforest was set aside for full government control.[40]

The Gujjars were once again alienated from their land. The British had placed a legislative stamp on a process they had started more subtly decades earlier: the assault on pastoral livelihoods and commonly held village land, and the replacement of such land with individual, salable plots interspersed with government-controlled enclaves.

From Sacred Grove to Real Estate

This process, which started on the Northern Ridge, took a long time to spread southwards; indeed, the story is still unfolding at the far reaches of the Southern Ridge and beyond. Here, one can find traces of older ways of life, of the "reservation of wood-producing land... generally connected with religion", which the British found when they first entered the region. The most remarkable example of this is the sacred grove of Mangarbani. The grove technically falls in Haryana, outside the bounds of the National Capital Territory of Delhi (a clear-cut boundary that owes its contours to the British). Officially it is not part of the Delhi Ridge. But it is on the Aravallis, so its underlying stones and soil are identical to those of the Ridge.

What grows out of this stony soil has amazed observers from Delhi. Mangarbani is densely forested, and it is full of trees that have long disappeared from the Delhi Ridge, most notably the graceful, hardy *dhau* tree. The thriving forest is part of common lands shared by the three villages of Mangar, Bandhwari and Baliawas, which are dominated by the Gujjar community; these villages maintain Mangarbani as a sacred grove in honor of a holy man named Gudariya Baba who used to roam these parts.

The very existence of such a remarkable ecosystem in the midst of three Gujjar settlements neatly refutes the argument originally put forward by Maconachie and Beadon, and still echoed today by some government officials and environmentalists: that villagers, especially those with strong pastoral traditions, like the Gujjars, do not care about trees, and land must be taken away from them and vested in the government if it is to be protected.

Instead, Mangarbani suggests another lesson: that the ecological behavior of pastoralists is molded by the larger systems in which they are ensconced. For example, during Mughal times, the Gujjars living close to Shahjahanabad put excessive pressure on their commons because of urban demands for meat and milk; meanwhile, the Gujjars of Mangarbani, much further away from city influences, left a more balanced ecological footprint. This remained the case during British rule as well, despite the compulsions of British land settlement efforts and despite pockets of quarrying. But, over the past several decades, Mangarbani has increasingly been drawn into the orbit of a rapidly expanding Delhi, a trend that has threatened the survival of the grove.

Regular readers of Delhi newspapers may be familiar with this story. On almost a weekly basis, one reads about a new threat to the old forest. Or about threats to people *in* the forest. By far the most dramatic headline came on 31 March 2014: "Birdwatchers Thrashed at Mangar Forest".[41]

The events described in the article are a combination of the disturbing, the tragic, and the absurd, a mix that characterizes much of the Delhi region in an era of runaway growth. The birdwatchers were from nearby Gurgaon—not too long ago, a sleepy farming village and now an urban hub conjoined to Delhi and filled with automobile factories, multi-national corporate offices, small-scale garment industries, and a seemingly endless expanse of malls. (The city was renamed Gurugram in 2016, a Hindu nationalist nod to the Mahabharata sage Guru Dronacharya, who supposedly lived in this area thousands of years ago. However, for the sake of avoiding anachronisms, I will stick with "Gurgaon", the name that was current when the following events took place.)

The birders had gone to Mangarbani to spot winged wildlife in the native forests of the grove. When the first car reached the grove, they came upon a man who said he was the priest at the local temple; he wanted to know what they were doing there. Things got a bit heated, and the priest took out his phone and made a call. Within minutes, a group of young men sped onto the scene in a jeep. They jumped down, armed with sticks and iron rods, and attacked the birdwatchers, a group which included an elderly woman and a young child. The attackers fled, though, when the rest of the birdwatchers, another four or five carloads, arrived on the scene.

The priest was later arrested, along with some of the assailants. Now, though, they are all out on bail, as the rusty machinery of the justice system does its agonizingly slow work. Many of the news reports after the attack asserted that the priest has played a central role in real estate transactions in the area. The British may have been the first to introduce the idea of land as a commodity in the Delhi region, but now, centuries later, the idea has become common sense. It is embraced with gusto by the wide range of players that make the real estate industry tick, a group that, apparently, includes a temple priest and his hired muscle.

Real estate is now the shadow that hovers, unavoidably, over Mangarbani and the three villages that surround it. This, though, is a relatively recent development, and it has gained traction due to the changing role of the Gujjar landowners in the villages. The fact that Gujjars are the dominant landowners suggests that, in this area at least, they long ago made the transition from nomadic tribe to settled community. Pastoralism still plays a role here, but it has long been complemented by agriculture, and it has taken place around fixed village settlements. And Gujjars have integrated into a caste-based village structure, finding themselves in a powerful position within the local hierarchy.

The complexity of the caste system is in full view with the Gujjar community. In most states in India, Gujjars come under the administrative category of Other Backwards Classes (OBC), which puts them below the traditionally "high" castes, but above Dalits (administratively: Scheduled Castes or "SC") and tribals (Scheduled Tribes or "ST"). It also makes them eligible for a range of reservations made available by the state. But this cut-and-dry state-imposed category hardly gets at the nuances and the internal differences within Gujjar communities. In some parts of India, especially in the Himalayan foothills, Gujjars still live a more tribal, nomadic existence, with little integration into settled caste systems; however, in other contexts, including Mangarbani, they are not only integrated, they are also the most powerful community in a given village.

While OBC may, then, be a wholly inadequate way to describe Gujjars, the designation is still vitally important, given its link to reservations. In some cases, Gujjars have demanded a lower status, so that they have access to more state benefits. These are the exigencies of modern-day caste politics. While the impulse behind reservations is a deeply progressive one—to provide support and opportunities to groups that have historically been exploited and marginalized— their application must deal with the messy terrain of competing communities, internal discord, and intersecting layers of privilege and power.

Such complex dynamics often lead to explosive results. In 2007, in the state of Rajasthan, a group of Gujjars began to agitate for the inclusion of Gujjars as a Scheduled Tribe (ST), in a sense a step "down" from OBC, but one which would provide them with more state support. As the protests gained momentum, they triggered state repression. Within a span of four days in May 2007, police opened fire on four different groups of protesters, in conflicts which left 25 Gujjars and one policeman dead.

In early June, the protest turned national, as Gujjar groups from around the country descended on Delhi and other major cities, including Jaipur and Ahmedabad. In a remarkable show of com-munity strength, the protesters successfully cut off all road access to Delhi, effectively blockading the national capital. While the agi-tation was largely non-violent, some protesters set fire to buses and trains. For the elite of Delhi, this destruction of property and inter-ruption of their everyday life could not be countenanced. All the slurs and all the urban disdain toward Gujjars, from Babur to the British, were dredged up. A municipal councilor in Delhi is on the record saying that, for Gujjars, "killing is in their blood". Protests conti-nued the next year, with 38 more Gujjars shot down by police. The agitation only stopped when the government agreed to give Gujjars reservations, not as a Scheduled Tribe, but as a Denotified Tribe— the official post-colonial term for groups that the British had dubbed Criminal Tribes.

At the height of the agitation, protesters from Rajasthan got strong support from leaders of Gujjar-dominated villages in the Delhi region, who both benefit and suffer from their proximity to state power. Their strength at the village level gives them significant pull in local elections, but despite that they cannot compete with the real power players of the capital. Economically, too, they have benefited from the ever-expanding markets of Delhi, but, except in rare cases, they have not found a place at the table of the city's elite.

The traditional ruling classes in Delhi still see Gujjar-dominated areas as a backwards hinterland, even though, with the expansion of the capital, they are often right in the midst of the urban sprawl. If not physically, they are still metaphorically on the edge of an urban zone that houses a far more powerful set of elites. And it is increasingly not just an Indian elite housed in the Indian capital, but an international elite housed in the multinational offices and luxury high-rises of Gurgaon. This is the larger context in which the Mangarbani drama has played out, as the sacred grove is being inexorably pulled into the capital's sphere of influence.

Interacting with these larger factors were village-specific dynamics. As the main landed caste in the three villages surrounding Mangarbani, Gujjars controlled the commons, including the sacred grove. Dalits and other landless families could only use the commons with the permission of the landowners. This began to change in the late 1970s, when the government of Punjab (of which Haryana was then a part) began a half-hearted land reform effort, carving out portions of the village commons to give to Dalit families as agricultural plots.[42]

The plan succeeded in making the government look progressive while leading to a series of regressive changes on the ground. The government's logic was that landowners would not be too opposed to these efforts because they were only taking away "waste" land— that is, the commons. This is in contrast to truly radical land reforms, which would redistribute fertile agricultural land and give everyone the right to access the commons. Such reforms would require a strong commitment from the state, which it was hardly prepared to give.

Breaking up the commons to give plots to Dalits was a ruinous plan for several reasons. First, especially in hilly areas like Mangarbani, the commons were hardly suitable for agriculture, so giving them to Dalits and telling them to farm there was something of an insult. Second, the Dalits themselves depended on the commons for fuel and fodder (even if their access to it was constrained), so breaking them up would hardly be in their long-term interest. Third, the move increased tensions between Dalits and Gujjars, as any land reform effort would do, but unlike effective reforms, it did not actually give Dalits the economic strength, or the state backing, to hold their own and fight for an egalitarian village structure. Instead, the move encouraged a backlash on the part of Gujjars, who felt (not without reason) that this was just one more move from on high to take their land away.

Not long after the government introduced the reforms, landowning communities from across the state banded together to change the law regarding village commons. They succeeded in the effort, and the government amended the law in question to allow common lands to be partitioned, and for each of the landowners to receive a portion of the commons commensurate to his existing landholding. In the end, the commons were broken up; but instead of being given to landless Dalits, they were hoarded by those who already owned land. Ironically, it was not the avarice of British land settlement policies that finally destroyed the commons of Mangarbani; it was the ersatz progressivism of the postcolonial state.

But the effects of the partition of Mangarbani's commons were not immediately evident. The initial partition took place in late 1970s, but it remained purely notional; it was still used as common land under the firm control of Gujjar village councils. However, in 1986, the individual plots were actually marked out on the ground. It is no coincidence that this happened just as Gurgaon was beginning to take off. Soon after the demarcation of plots, real estate agents and speculators descended on the villages, and many villagers sold off their land. Several of the plots have now been fenced off, in preparation for building when the land values become appropriately lucrative. So far, this has not happened in the sacred grove itself, but only around the edges.

Villagers selling their land, and potentially abandoning their sacred grove, should not be read simply as a crumbling of traditional ethics and a sign of our degraded times. These decisions must be placed in a larger economic context. Pastoralism has long ceased to be a profitable activity; the British started to kill the commons and the open grazing routes in the region, and the postcolonial government finished them off. From the 1980s onwards, when Delhi was in the throes of a building boom, mining extended to the far reaches of the Aravallis, including many of the areas surrounding Mangarbani. This propped up the local economy; the total ban on mining in 2002 (largely in the name of environmental conservation) caused widespread economic uncertainty in the zone. In this situation, many Gujjars, who had been devoted caretakers of the sacred grove, felt they were better off selling their share of it.

Greed likely played a role, as did a lust for power; how else to explain a priest so zealously taking up the role of real estate broker and violent enforcer? But even this must be seen in the context of the larger forces swirling around the region. The liberalization of the economy, which had its watershed moment in 1991 and which has Gurgaon ("The Millennial City") as its iconic location, brought with it a wave of consumerism, an unapologetic celebration of the accumulation of wealth, and a reveling in newly-available imported products. Many of the younger generation in villages around Mangarbani were already drawn to the city out of economic compulsion. Is it unreasonable that they were swayed by the new ideology of consumption that was so openly celebrated by the city elite?

This question will be considered in more depth in Chapter Four. For now, it's important to note that some were not convinced by the new gospel of spending. Many from the villages near Mangarbani, especially but not exclusively the elders, still want to hold on to their sacred grove. Some who sold their land now bitterly regret it. There was a wave of excitement when plots were first being bought for sums of money that then seemed unimaginable. But the real profits, it eventually became clear, were made by corrupt speculators and real estate moguls selling and re-selling the land.

The real estate industry is well known for chicanery and its collusion with state officials (a full exploration of this nexus will once again have to wait until Chapter Four). In the case of Mangarbani and other Aravalli areas in Haryana, government officials have misclassified hilly zones, which for centuries have been used for grazing, as agricultural. Once an area is categorized as agricultural, it is relatively easy to convert it into a commercial or residential zone, especially as huge waves of urbanization pass through the region and turn even legitimate farming areas into city neighborhoods and suburban getaways. Real estate moguls find out about these zoning changes in advance (and, in some cases, actually make these changes happen), and can reap huge rewards as land prices rise astronomically.

Questionable zoning is just one part of the real estate-government nexus. The public learned more about this shadowy alliance thanks to the revelations of Ashok Khemka, a bureaucrat and whistleblower famous for being transferred 45 times (and counting) in his 22-year career. His uprightness and zeal for countering corruption have invariably created tension with his supervisors; many of his postings have lasted only a few months. In 2012, Khemka canceled several plans to consolidate landholdings in the area around Mangarbani, which was hilly, uninhabited and tree-covered, if not as lush and well-protected as the actual core of the sacred grove. A group of real estate agents had been buying up low-priced plots of land, and then trying to illegally exchange those plots for much more valuable ones in areas where new roads were coming up. These agents often represented high-level politicians and businessmen who knew about these roads when they were in the early planning stages and could conveniently invest their black money in a foolproof moneymaking scheme. Although Khemka interrupted some of these deals before his inevitable transfer, he could hardly put a dent in the larger corporate-political machine that is chewing up Aravalli forests and spitting out luxury housing and commercial complexes.[43]

In spite of such threats, the grove at Mangarbani is still standing.[44] An alliance of city environmentalists and village advocates have, after a long battle, succeeded in getting the area officially demarcated as forest land. But its long-term prospects are uncertain. Mangarbani is now unmistakably in the orbit of urban Delhi, and the Gujjars of the surrounding villages must now face a dilemma that the Gujjars of Delhi faced long ago: how to deal with an immensely powerful political and economic elite hungry for land and money. Some villagers near Mangarbani have taken an active role in facilitating the ongoing land grab, and many have been willing to sell their land. Others have stood firm and resisted the onslaught. But even they can sense that a massive change is coming.

The ongoing saga of Mangarbani aptly illustrates the ever-changing ecology of the Ridge, as it transforms from forest to grazing land to mines to real estate. Throughout history, the Gujjars have shaped the Ridge, and the Ridge has shaped the Gujjars, with the burgeoning mass of urban Delhi increasingly deforming and reforming both.

Migration of Gujjars

Likely arriving from Central Asia, Gujjars move throughout northern India, as pastoralists, warriors, agriculturists.

Some become rulers, while others are displaced.

Their commons are converted to "public" property.

Many Delhi quarries are sited on their lands.

Some, like Mangar Bani, remain, but face threats.

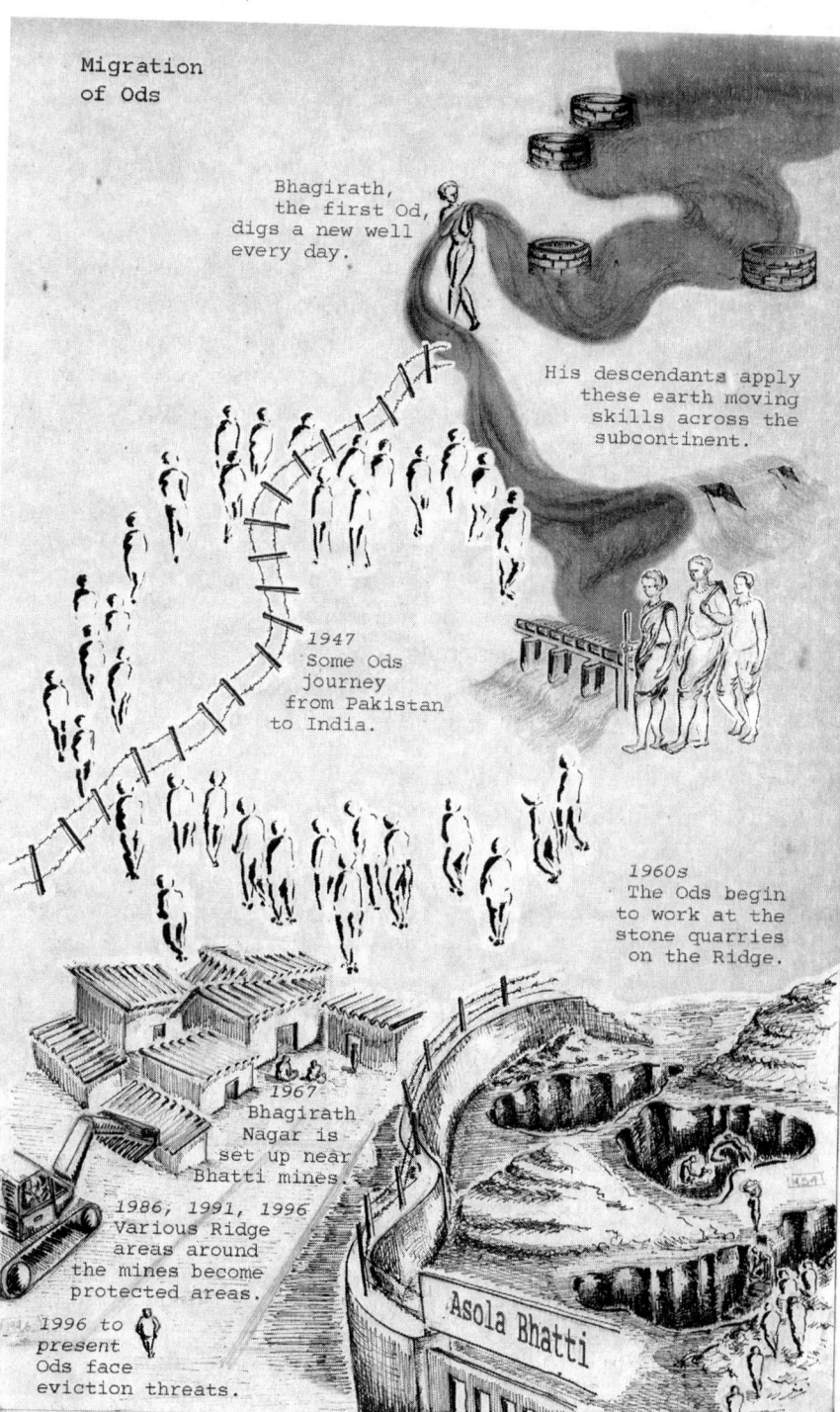

Migration of Ods

Bhagirath, the first Od, digs a new well every day.

His descendants apply these earth moving skills across the subcontinent.

1947
Some Ods journey from Pakistan to India.

1960s
The Ods begin to work at the stone quarries on the Ridge.

1967
Bhagirath Nagar is set up near Bhatti mines.

1986, 1991, 1996
Various Ridge areas around the mines become protected areas.

1996 to present
Ods face eviction threats.

Asola Bhatti

The transition from a nomadic to a settled life, due to the twin forces of British land settlement policies and urban expansion, was a radical, but likely quite gradual transformation of Gujjar life. While their connection with the Ridge was perhaps most vivid and vital when they moved through it as nomads, their political and social claims to the land grew stronger with their entry into the ranks of settled gentry; they became "sons of the soil" in a fairly traditional sense. Now, the move to turn the Ridge into a vast real estate market is destroying the gentry as a class but is providing other avenues for Gujjars to accumulate power and profit. Their fate then, continues to be bound up with the changing ecologies and economies of the Ridge.

Other Nomadic Trajectories

The Gujjars may be the most iconic inhabitants of the Ridge, and the ones with the farthest-reaching historical claims. But it would be a mistake to see them as the only notable inhabitants. Their dominant role should not obscure other groups who have traced out similarly complex passages from nomadic tribalism to settled village life, and from pastoral pursuits to the exigencies of urban living.

One such group is the Ods, traditionally a nomadic community who, like the Gujjars, are spread across much of northern India. Also like the Gujjars, the Ods trace their lineage to ancient royalty and also consider themselves to be Rajputs. However, while Gujjar history can be traced at least in rough outline from the seventh century onwards, Ods appear more suddenly in the historical record. Ods have long been spread throughout northwest India, but the group of Ods now living in the Delhi Ridge only entered the area after Partition; before that, they were in Sindh, in what is now Pakistan.

Unlike other migrants from the northwest, their choice of animal companion was not the fighting horse, but the humble donkey, useful for carrying loads. Not known for their agricultural prowess, Ods are sons of the soil in a different sense: they are diggers of the earth, constructors of wells, ponds, and embankments. This profession has mythical origins: an ancestor of the Ods, Bhagirath, once declared that he would never drink water more than once from the same well. He thus condemned himself to a life of digging a new well every day, which had the unintended consequence of honing his earth masonry skills considerably.

This set of skills has found plenty of application in modern, post-colonial India; Ods have scaled up their innovative digging techniques from individual wells to the mega-projects of the independent state. Od community leaders in Delhi reel off the list of big-name projects for which they helped lay the foundations, from the Indira Gandhi Canal that stretches from Punjab to Rajasthan, to the airport and renovated railway stations in Delhi, to the Badarpur thermal power plant, to the All India Institute of Medical Sciences (AIIMS), to a series of dams, including the Tehri Dam and the Bhakra Dam. Nehru famously described dams as "temples of modernity", but many of them were built with the traditional know-how and earth-moving skills of Od workers.[45]

It is precisely their facility with soil that brought Ods to Delhi in the 1960s, and specifically to the Bhatti mines, the largest quarrying complex of the Delhi mining industry. Its main product was "Badarpur", the red sand created by the erosion of iron-rich quartzite, although uneroded quartzite blocks were also quarried. With their digging skills and intuitive sense of the earth, developed over many generations, Ods were well-suited to mining work, and they quickly became the largest community of workers in the Bhatti mines.

When Ods tried to set up a more permanent base near the mines, they encountered resistance from the settled Gujjar community that dominated Bhatti village, as well as the surrounding villages. But the Bhatti mines were feeding the booming construction industry in the capital city, and the mines needed workers, so Ods quickly found political backing, at a remarkably high level. Indira Gandhi, then the prime minister, set up an "Od Nomadic Tribe Cell", which supported the establishment of schools, health centers and veterinary clinics (for the donkeys). A foundation stone still stands for a housing project inaugurated by Indira's son, Sanjay Gandhi, in 1976; after Sanjay's death in 1980, the main Od settlement in the Bhatti mines changed its name from Bhagirath Nagar to Sanjay Colony.

The Ods were also beneficiaries of half-hearted land reform efforts, similar to the ones that marked the beginning of the end of the Mangarbani commons. In 1987, many Od families received plots of land that had been carved out from the village commons, in a move that clearly rankled the local Gujjars. The official record of these transactions has been lost; many Ods believe that this is the doing of local Gujjar administrators, who held most of the lower-level government positions in the area and who had ample reason not to record the new land rights of the Ods.

But, as with many of these relationships, it was not simply one of strife; after the closure of the Bhatti mines, Ods and Gujjars both found themselves facing economic catastrophe, and they lobbied together to have the mines re-opened. For Ods and other mine workers, though, the crisis struck not just their livelihoods, but their homes. Sanjay Colony and two smaller workers' settlements were suddenly labeled 'slums', which were to be cleared in order to extend the Asola Wildlife Sanctuary. Enlarging this sanctuary, which had first been established in 1986, and renaming it Asola-Bhatti was a convenient way to cover up the mess of the mines. Bhatti had been the site of the most egregious labor abuses and deadly accidents in the quarrying belt. In 1996, some Od villagers were watching TV and learned that their houses were slated for eviction; the state did not bother to give them an official notice.

Since then, Od villagers have come to depend on the protection of powerful Gujjar politicians. Stripped of their livelihoods, with their habitation constantly threatened, Ods have little power except for their numbers. They leverage this power by supporting politicians who can make their lives marginally more secure. From 1996 to 2004, they found a tolerable patron in Chaudhary Brahm Singh Tanwar, the MLA of their constituency and a member of the BJP; his last name is enough to identify him as a Gujjar, and a member of the clan that ties itself to the Tomars, founders of Delhi. Tanwar made sure that the Od villagers got water from tubewells and received metered electricity connections, and he implemented a series of welfare schemes. His successor, another Tanwar, but from the rival Congress party, was not as generous, souring the Ods against the party that had originally supported them.

In 2006, the initial order to demolish the three Bhatti Mines settlements was finally put into action. On 20 April, the two smaller villages were overrun by police, who cordoned off the area and made way for bulldozers to do their sickening work. Afterwards, the villagers moved quietly through the wreckage, salvaging whatever could be taken to rebuild their lives. An older man died of a heart attack while trying to pull out a beam from his demolished house. Women gathered and started wailing, as government officials ignored it all and went about their business.

When the police arrived at the main settlement of Sanjay Colony, however, they could not proceed so easily. The colony was six times as large as the other two, and the demolition was initially planned as a multi-day process. But the would-be demolishers were met with an angry crowd, as almost all the colony's residents—men, women and children—rushed out to block the road and resist the police. They succeeded in driving them away, and they were similarly successful the next day. They got some relief soon after, when the Delhi High Court served a contempt of court notice to the Delhi government, criticizing it for its failure to submit an adequate relocation and rehabilitation plan for the villagers.

Now, well over a decade later later, Sanjay Colony still stands, but it is haunted by the looming threat of demolition. Various court orders have come in, upholding the demolition order, and various last-minute pleas and protests have pushed the demolition further into the future. The Od villagers are deeply disillusioned, bitter that the state that extracted so much labor from them, that relied on them to build the landmarks of the city, is now denying them any recognition. They have abandoned the name Sanjay Colony, recalling as it does unmet government promises, and gone back to the old community name of Bhagirath Nagar.

Much of their anger is now directed at the Forest Department, which runs the Wildlife Sanctuary that borders their village. For many years after its extension into Bhatti, the "wildlife sanctuary" was anything but; it was filled more with trucks than with animals. The sanctuary is at the very edge of Delhi, on the border of Haryana, and many trucks carrying materials from the still-active quarries in that state would roll through the sanctuary on the way to deliver their goods to the capital. The land remained denuded and degraded, full of the dust and smoke of the heavy vehicles. The situation was so dire that the government called in the army, or one part of it anyway, the army's Eco-Task Force. Such task forces had been formed all over the country as a way to give jobs to servicemen done with their tours of duty. This strategy represents an extreme version of the forest policies introduced by the British: take the forests by force and guard their borders vigilantly.

With their guns and their uniforms, the Eco-Task Force patrolled the perimeters of the erstwhile Bhatti Mines, and succeeded in keeping the trucks out. But their other mission, reforestation, has been less of a clear-cut success. In the early years of their reforestation efforts, the army task force mainly planted a Mexican mesquite species known locally as *vilayati* (foreign) *kikar*. This species was originally introduced by the British in their efforts to reforest the Ridge, after their attempts to plant other species had failed, due to the rocky soil and the aridity of the climate. *Vilayati kikar* has thrived on the Ridge, but at the expense of other species; its leaves contain germination inhibitors, which limit the growth of grasses, herbs, and small trees around it.

The Eco-Task Force chose the mesquite precisely because of its weed-like tenacity; it spreads quickly, with little effort. In their own way, the postcolonial government has shown the same ecological blindness as their colonial predecessors; the landscape is seen instrumentally, as a blank space in which to cram as many trees as possible (whether to be used for railway construction and so on, or to create the image of a "clean and green" city). The *vilayati kikar* has transformed the landscape into a monoculture thicket, far from the diverse tree-and-grass-filled savanna that the Ridge once was.

In recent years, in the face of pressure from environmentalists, the Eco-Task Force has stopped planting *vilayati kikar*, and has begun planting local species instead. But the damage has been done, largely due to the mesquite's tenacity; it has firmly established itself as the dominant species in much of the sanctuary. The Od villagers are well aware of this; they know all about the tree and its knack for pushing away all other species. But what really galls them is the way that the façade of green regeneration has been used to keep them in a constant state of insecurity.

The latest offensive against the Ods has shifted gears, from the ecological to the religious; the Ods are not just tarred as environmental encroachers, they are conflated with the Muslim Other. A senior forest official (with intelligence from the army's Eco-Task Force, no less) is claiming that Bhagirath Nagar is becoming a "Pakistani mohalla". His assertion is that "foreign nationals" from Pakistan are migrating across the border and finding their way to Bhagirath Nagar, although he admits that this is based on "unconfirmed reports".[46] His report implies that these are extended family members of the Ods who settled in the colony decades earlier. It does not matter that the Ods have vehemently denied this allegation, nor that the Ods are Hindu. Though the forest official never explicitly uses the "M" word, his rhetoric is a clear reflection of the all-too-common identification of Pakistan with Islam with terrorism. He warns that this is an urgent matter of "national security", which necessitates "surveillance" so that dire consequences can be avoided.

The Ods, though they are resisting this rhetoric, know they are facing the risk of a forced return to their nomadic lifestyle, if the state has its way and finally proceeds with demolitions. Perhaps it is because of their precarious situation that they are staking a claim to the Ridge's forest in a way that Gujjars have not. Gujjars, long settled, long having taken up the mantle of "Hindu resisters to Muslim invaders", now have little use for the mesquite-dominated hills that once yielded mountains of construction material. That economic opportunity, dominated by Gujjar contractors, has now closed. New ones have opened up, notably in the growth of palatial farmhouses around the sanctuary. The Ods, in their insecurity, depend more on the forest: for the firewood they need to cook and for the fodder to feed their animals.

Even though their village still remains standing, many Ods have effectively become nomads again, as they travel throughout the city and through the larger region searching for work. Their digging skills and their facility with the earth is still appreciated; in recent years, they have done a significant amount of work installing underground Internet cables. From mega-dams to high-speed Internet, Ods have laid the foundations for the newest technologies and most prestigious projects in postcolonial India. For that, they have had their identity questioned, their villages destroyed, and their liveli-hoods made illegal.

A Closer Look at Bhagirath Nagar

Bus Number 523 runs from Dhaula Kuan, a neighborhood bordering the Central Ridge, to Bhagirath Nagar. The official destination, though, is "Bhatti Mines". This flashes on the electronic yellow signboards of the bus, even though the mines have closed decades ago. As the bus approaches its destination, it turns off the crowded Mehrauli-Gurgaon road, through the luxury farmhouse zone of Chhatarpur, to a landscape increasingly defined by huge fences, erected by the government to demarcate the Southern Ridge. At the end of a dusty, bumpy dirt road is Bhagirath Nagar.

The road was once paved, during the heyday of government support when the mines were still churning out sand and stone; the courts have now forbidden the local government from repaving the road, as it will "encourage encroachment". Bus 523 runs infre-quently; it is supplemented by vans that serve as shared taxis, people piling in as they return home from work. They often carry the tools of their trade home with them; pickaxes, spades, shovels.

At the entrance to the village is a small chai shop (a brick structure with a thatch roof) and a temple (a more elaborate affair), and then a road with neat rows of houses on both sides. Moving down the road, one gets a sense of what the village may have been like, and what it has become. Despite the many government attempts to label Bhagirath Nagar as a run-of-the-mill "slum", it feels far removed from the crowded working-class neighborhoods elsewhere in Delhi. There are still pockets of space, still rural reminders, as mud houses are interspersed with newer concrete ones. Along the road are a few shops—a tailor, a vegetable vendor, a mobile repair shop. Pigs, donkeys and cows pass by.

On the left is a large concrete building behind a high fence. Outside, there's a sign with text in Devanagari script, but the words it spells out are English ones: "Indian Population Project". This was, according to the residents, a World Bank funded project, aimed at promoting family planning. The building is decrepit; windows broken, debris gathered in the courtyard. It has shut down, as have the other health care facilities in the area. Village leaders recount stories of women dying during childbirth because of the lack of facilities.

Not far from here is an upmarket retreat center owned by the Radha Soami Satsang Beas, a spiritual organization with deep pockets. It was built after leveling a hill, clear evidence that this was part of the Ridge, but the construction was allowed: it didn't hurt, the villagers note, that the Lieutenant Governor of Delhi at the time was a devotee of the sect.

No such favors, of course, have been extended to the residents of Bhagirath Nagar, at least since the mines were closed. Government attempts to destroy the colony have failed, due to the militant resistance of the community, so the state is attempting a siege instead, withdrawing welfare services and withholding employment opportunities. Many of the buildings along the main road are showing the strain; some are completely dilapidated.

The main road is short, and it ends abruptly, with an imposing gate. On either side of the gate are huge walls, roughly 20 feet high, marking the boundary of the Wildlife Sanctuary. The gate itself, though, is shorter, around seven feet. Easily scalable. Cross to the other side and enter into another world.

Border-Crossing Species

The Wildlife Sanctuary is both eerily silent and oddly posh. The dirt roads are neat, flat and clean, carefully maintained. The trees are aligned in a grid-like pattern, evidence of a planting effort that emphasized geometric precision.

But for all its apparent order, the Sanctuary is still full of surprises. It is, in a sense, an attempt to recreate the supposedly "pristine" nature of the Ridge, which for the Forest Department means keeping humans out. But the species that proliferate in the Sanctuary straddle the boundary between nature and culture, between wilderness and human influence. They mock attempts to erect clear boundary lines. This is especially true of the two most prominent species of the Sanctuary, one of which dominates the flora and one of which dominates the fauna. The former is not too difficult to guess: the ever-present *vilayati kikar*. The latter may not be a surprise either, at least for those who have visited other parks in Delhi: the rhesus macaque, that ubiquitous urban monkey.

These are not the "native" species that environmentalists prize. And yet they tell us far more about the present-day Ridge than indigenous plants or animals that have not existed there for decades. The modern Ridge forest is, after all, of relatively recent vintage, and it is in the middle of an ever-expanding megacity. It abounds in hybrids and muddled categories, in odd mixes of the city, the countryside and the jungle.

The environmentalist's disdain for *vilayati kikar* is not without basis. Besides the fact that it was brought by the British, who demonstrated their arrogant misunderstanding of Indian ecology again and again, it has clear ecological downsides. And yet the intensity of the scorn heaped on the mesquite seems disproportionate to its ecological crimes. It is not as if it pried its way into a thriving native forest, strangled all the indigenous inhabitants, and then put down its own thirsty roots. Yet this is exactly what is suggested by references to *vilayati kikar* as an "aggressive colonizer" running rampant on the Ridge. In truth, the tree was brought into a heavily degraded landscape, pockmarked by decades of mining.

Yes, it would have been far better if the Forest Department and their friends in the army had been more sensitive, more forward-thinking, more patient. Yes, they could have allowed a more diverse ecosystem to grow back over time, instead of going with the quick fix of *vilayati kikar*. But a sea of *vilayati kikar* is the landscape now confronting us. However ecologically compromised, the green cover of mesquite is still a green cover, with all the benefits that this implies: cooler temperatures, more oxygen, a first line of defense against the dangerously high levels of air pollution in the city.

There are more localized benefits as well: the residents of Bhagirath Nagar use the branches of the *vilayati kikar* extensively for firewood. Back in its native Mexico, as well as in the United States, mesquite is known for the delicious, smoky flavor it imparts to food cooked on it. This is an incidental bonus for the Ods; they mainly collect it because it's the most easily accessible fuel source. Even here, though, the Ods don't have it easy, since collecting firewood in a Reserved Forest is illegal. Usually women are given the task of gathering the wood, and all too often, they are harassed and intimidated by Forest Department officials and army tree-planters.

By now, even the Forest Department has realized that *vilayati kikar* was not the best ecological choice; and yet government officials have so internalized the notion that the Ods are "outsiders" and "encroachers" that they take the side of the recently-planted, thirsty tree over the long-suffering Ods. With a modicum of sensitivity and creativity, a mutually-beneficial solution could be reached: Ods could employ their formidable skills to uproot the tenacious mesquite, use the tree for firewood, and help the Forest Department in its effort to plant more appropriate species. This would generate employment and help the ecology of the region. The impediments to such a plan are not logistical, but political. The Forest Department is loath to give up its control of the Sanctuary, even if that just means letting more locals in to work. And the government doesn't want to undermine the picture it has painted, of Ods as outside invaders who must be pushed aside.

So now both Ods and *vilayati kikar* trees are demonized by those who want a "clean and green" Wildlife Sanctuary full of thriving native species. This is not, though, the attitude of the city as a whole, at least with respect to the trees. While most ecologists still emphasize the "vilayati" part of *vilayati kikar*, and enumerate its negative qualities, it has, in common parlance, become known simply as *kikar*; the term is used interchangeably for both the local and the "vilayati" variety. The citizens of Delhi have, through their everyday vocabulary, naturalized the foreign tree, a city of migrants easily accepting one of its own.[47]

Now a naturalized citizen, the *vilayati kikar* at least has some limited benefits, especially for the put-upon residents of Bhagirath Nagar. The monkeys, though, are a different story. Media outlets have coined the term "monkey menace" and for once, they are not exaggerating. Nearly every resident has a monkey story, and they are not pleasant tales. Monkeys biting children; monkeys stealing precious supplies of rice and wheat; monkeys grabbing laundry and

Migrations

Rhesus Macaques

Rift Valley

Aryans

Central Asia

Vilayati keekad

1800s from Mexico

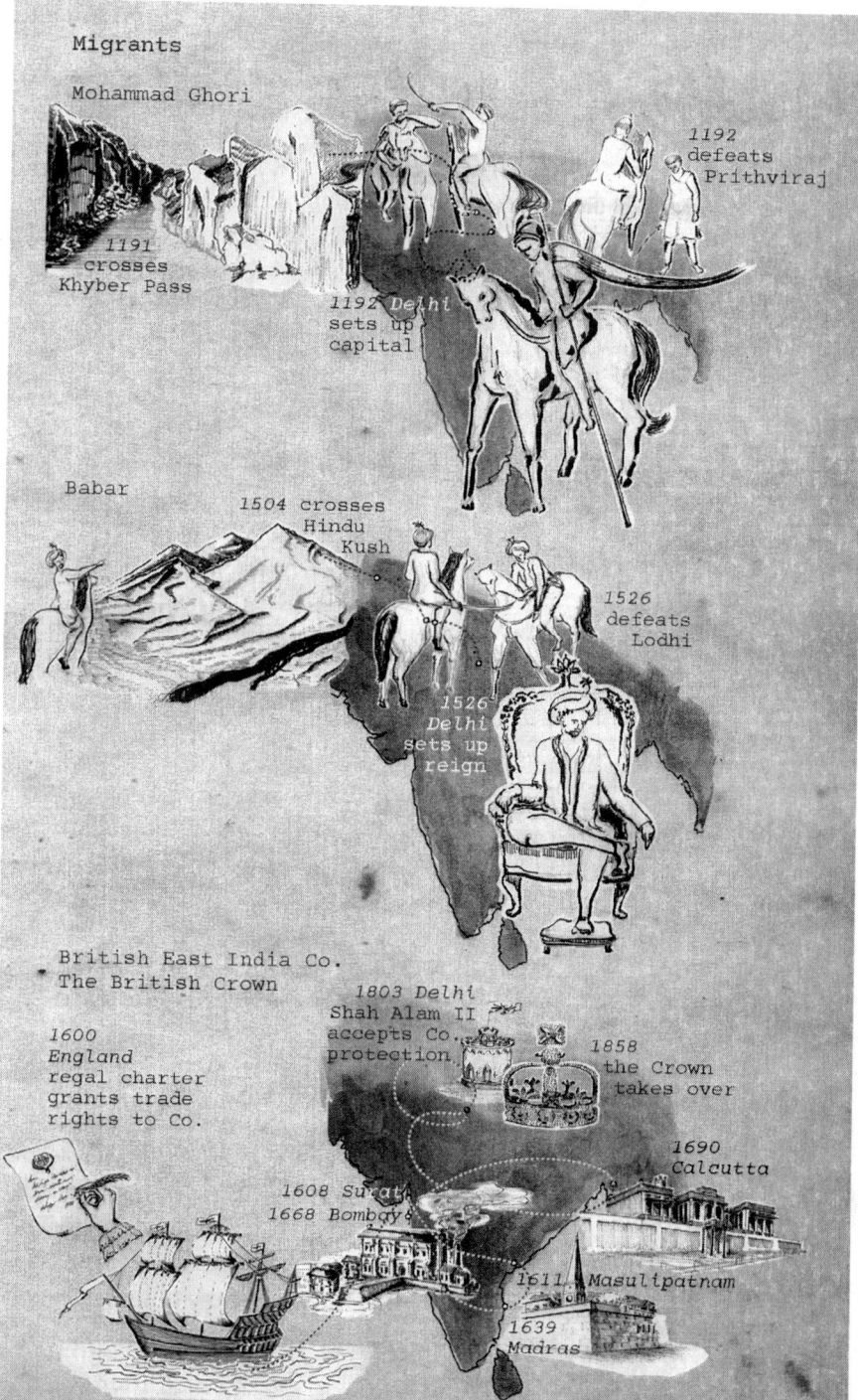

Migrants

Mohammad Ghori

1191 crosses Khyber Pass

1192 Delhi sets up capital

1192 defeats Prithviraj

Babar

1504 crosses Hindu Kush

1526 defeats Lodhi

1526 Delhi sets up reign

British East India Co.
The British Crown

1600 England regal charter grants trade rights to Co.

1803 Delhi Shah Alam II accepts Co. protection

1858 the Crown takes over

1690 Calcutta

1608 Surat
1668 Bombay

1611 Masulipatnam

1639 Madras

tearing up clothes. Monkeys can carry rabies, and the villagers know they should get an anti-rabies injection if they are bitten, but they can hardly afford it; one shot costs ₹500, and a full course of injections is five shots.

But the monkey, unlike the mesquite, cannot simply be dismissed as a rank outsider. Monkeys roamed the Delhi region long before the first walled city arose. What is more, Delhi has a schizophrenic relationship with its monkeys; it fears them and hates them for their destructive tendencies, and yet reveres them as manifestations of the god Hanuman. The city feeds the monkeys as a religious duty, then complains when the monkeys come looking for food.

The problem is not confined just to Sanjay Colony. In 2004, monkeys managed to gain access to the Defense Ministry and scatter papers about, in their typically destructive way. In 2007, the Deputy Mayor of Delhi died after falling from the balcony of his home in East Delhi while trying to fend off monkeys. This incident, at least, put the matter on the government's agenda, as did a case just a few weeks later in which a rampaging monkey bit 25 people over the course of two days in the East Delhi neighborhood of Shastri Park.

The mayor announced that measures needed to be taken. But which measures? The obvious first step was to increase the number of monkey-catchers; in 2007, there were only five government-employed monkey-catchers for a simian population of roughly 20,000. However, it was difficult to recruit monkey-catchers, since many saw the profession as sacrilege, an affront to Lord Hanuman. Still, suitable candidates were found, and a concerted effort to round up the monkeys began.

But where to put them? Here, the Delhi High Court intervened. Following the advice of animal activists, they suggested, why not put them in the Wildlife Sanctuary? This is when Bhagirath Nagar's monkey problem began. Although the Wildlife Sanctuary is huge, the government chose to erect the monkey resettlement zone on the small section of the sanctuary bordering the Od village. The metaphor is all too obvious: the government treats its unwanted animals like it treats its unwanted humans, forcibly ejecting them from their homes and pushing them to the furthest corners of the city.

The monkey's new home, like all government resettlement colonies, is bleak. The government has erected tall green plastic sheets to keep the monkeys enclosed in their area of the sanctuary, but this has hardly been an impediment to a species known for its agility and climbing prowess. The monkeys escape their prison whenever they feel the urge to do so. They have also started tearing down some of the plastic sheeting, a mute protest, it seems, to their new fate.

But some twinge of consciousness, some residual devotion to Hanuman, made the government feel at least marginally responsible for the monkeys they had displaced. The court ordered that two prominent Hanuman temples should start collecting food donations from devotees, and that these should be taken to the sanctuary on a regular basis. If this is not enough food for the 16,000 or so monkeys now in the sanctuary, then the government is responsible for making up the difference. According to a report by the environmental magazine *Down to Earth*, supplying monkey food at such high volumes has become a lucrative, extremely corrupt business.[48] The government spends more than one crore rupees a year feeding the monkeys, but the supply often stops unexpectedly because the forest department fails to pay the middlemen who have won suspiciously large contracts for monkey food delivery.

When there's no food, and even when there is food, the monkeys climb out of their prison and raid the food supplies of Bhagirath Nagar. They seem to have a particular liking for roti, but they'll take whatever they can get, biting anyone who tries to stop them. A doctor at a nearby clinic reports that he gets at least ten cases of monkey bites a day. Adding insult to injury, the Forest Department claims that the Bhagirath Nagar residents are actually stealing the monkeys' food. Perhaps this is to distract from the fact that the department has failed in its responsibility of planting fruit trees that were supposed to provide the bulk of the monkey's food.

The faulty prison, with its questionable food supply, is clearly just a stop-gap measure that doesn't actually address the root causes of the monkey problem, and instead just dumps the problem at the doorstep of a group that the state has already written off. Just what are those root causes?

Shivam Vij, a journalist who has written two outstanding articles on the history of Delhi's monkey problem (and who himself has been harassed by roving bands of monkeys), has taken a historical perspective on the matter.[49] With the help of primatologist Iqbal Malik, Vij traces the "original sin" back to the 1920s, when American researchers started taking monkeys from India and bringing them back to the US for scientific studies, especially biomedical research. The monkey trade only increased after Independence, as the country found itself in dire need of foreign currency; it could make valuable dollars by exporting simians. At its peak, roughly 50,000 monkeys a year made the journey to the US. This continued until the practice was stopped in 1978 by the devout Prime Minister Morarji Desai, who was disturbed by the high numbers of monkeys who were dying en route or in the course of experiments.

But by then, the damage had been done. The researchers preferred male monkeys who were not too old and not too young. This preference, along with the sheer volume of the trade, severely disrupted monkey society. In her research in the 1980s, Malik found that the sex ratio of monkey populations in Delhi was skewed, as were the ages of the monkeys. The social life of monkeys is a highly involved one, and the practice of abducting thousands of middle-aged males broke down the bonds that kept monkey communities together. The remaining monkeys became more aggressive, and more desperate for food, and hence more willing to venture into human territory to steal rotis, snatch bananas and so on.

However, the matter does not rest here. The roots of the problem go much, much deeper, and they complicate the common-sense understanding of the monkey menace. The story is not just one of humans interfering with pristine nature through kidnappings and construction activities, disrupting the happy simian families that had done so well before our meddling. On the contrary, there is considerable evidence that this particular species of monkey has actually evolved in human-dominated environments and has developed traits that are advantageous for these settings.

This is the argument, at least, of a scientific study tellingly named "Weed Macaques: The Evolutionary Implications of Macaque Feeding Ecology".[50] Like *vilayati kikar*, macaques are "weeds" that thrive at the expense of other species. Scientific studies have made it clear that rhesus macaques do extremely well in human-influenced environments, and in many cases seem to prefer them. Of all the rhesus macaques living in northern India, roughly half the population is concentrated in cities, towns, temples and railway stations, that is, places with a lot of human traffic. So maybe, when it comes to Delhi, the story isn't just of humans invading monkey terrain, but of monkeys seeking out human terrain.

The rhesus macaques, after all, came quite a way to get here, especially if we take the long evolutionary view. Rhesus macaques are just one of about 20 macaque species. Like humans, macaques originated in Africa. The ancestor of all current macaque species probably emerged around the same time as the first hominids. Unlike humans, who likely came to India directly from Africa via the coast, macaques took a more circuitous path, and broke up into many different species along the way. The early macaques crossed land bridges into Europe, and from there they spread eastward, eventually entering Asia.

By this point, several macaque species had emerged, forming four large groups. The rhesus macaques belong to the last of the four groups, the one that split off from the other macaques most recently. Their entry into Asia is relatively recent. It has also been spectacularly successful. Rhesus macaques have a knack for pushing other macaque species out of their habitats, as well as for venturing into new places where no macaque has gone before. As a result, rhesus macaques are the second most widespread primate species in Asia, second only to human beings. It may not be a coincidence that these two species are at the top of the list. The rhesus macaque, as compared to other macaque varieties, is exceptionally comfortable around humans, bolstering the claim that the species has likely evolved in human-influenced environments.

This means that, in the course of its evolution, the rhesus macaque developed characteristics that helped it both expand to new locations and co-exist with humans. The primatologist Dario Maestripieri identifies three such characteristics: curiosity, aggression, and militaristic social formations. The first is the most benign, and the most in keeping with the monkeys that populate the world of cartoons, children's books and movies; there is, after all, an entire book series revolving around a monkey named Curious George. Rhesus macaques live up to that billing; they like trying out new foods, exploring new places, climbing new trees and scaling new buildings.

The flipside of this curiosity is aggression. Rhesus macaques, with their expansionist ways, are bound to run into new animals. When they do, they size up their potential foe and, if they think they have the upper hand, they attack. If no new animal is around, they are content with fighting each other. As Maestripieri says, "Rhesus macaques are irritable creatures who have a low threshold for aggression. They wouldn't make it in the world without their fellow macaques, but they have very few inhibitions against attacking and hurting one another."[51]

The aggressiveness of rhesus macaques is not just an individual trait; it is bound up with the way their society is organized. Rhesus macaque society is, Maestripieri tells us, intensely nepotistic and despotic. (He's even come up with a clever pun for this: "Macachiavellian".) It is a matrilineal society, with males leaving the community when they reach puberty to find another band of monkeys. Females are thus a more constant presence in the communities, and they form the core of the social structure. Having the females in charge doesn't make things any more peaceful, though. Each community is defined by a hierarchy of extended families, with the more powerful families engaging in constant acts of violent harassment to maintain their dominance in the hierarchy. Maestripieri compares this to the social organization in an army: everyone knows their rank, everyone shows deference to their superiors, and everyone communicates in a terse language geared towards relations of dominance.

And like an army, rhesus macaques are geared up to go to war. As a species, they are highly xenophobic; they don't like strangers, and they don't like other rhesus macaque communities. When one entire community goes into battle with another community, internal hierarchies and differences are put aside, and the group bands together against the outside threat. In fact, the low-ranking monkeys in the community use these battles as an opportunity to vent their frustration, and they often prove to be the most effective, vicious fighters.

When two rhesus macaque communities already know each other, and have already established which group is dominant, their battles are relatively tame, as they are just minor adjustments to an already-fixed order. The monkey fighters are rarely killed, or even seriously injured. However, when two totally new communities meet, the stage is set for a gruesome bloodbath.

Scientists learned this the hard way. One of the first major studies of rhesus macaques began in 1938, when an American biologist named Clarence Ray Carpenter traveled around north India collecting rhesus macaques, then threw them all together in a ship and sent them to a tiny island off the coast of Puerto Rico, where he could observe them at leisure in a laboratory-like environment. On the ship, though, there was vicious fighting, with several of the monkeys drowning after being forced overboard by their fellow simian passengers. Scores of infants were also killed. When the survivors finally reached the island, there was a final murderous battle, as the monkeys set out to establish their hierarchies and dominance relations once and for all. After this, things became (relatively) peaceful; there are four major clans on the island, and each knows its place.

This harrowing journey reveals something important about these monkeys; the chief threat to their survival is not, it seems, changing ecological conditions, but rather their fellow rhesus macaques. They do just as well in Puerto Rico as they do in Delhi. In fact, this is why they are so prized by biomedical researchers; they'll thrive in any laboratory, anywhere in the world. But they may not survive an interaction with the monkey clan that lives next door.

This may explain why the monkeys in Bhagirath Nagar are so surly, why monkey bites are so common, and why the monkeys are so eager to escape their cage in Asola-Bhatti Wildlife Sanctuary. The government has hired monkey-catchers to gather a bunch of xenophobic, aggressive, irritable macaques from different Delhi neighborhoods and to throw them all together in one location. It's no surprise that the place has become a war zone. The villagers of Bhagirath Nagar are just collateral damage.

The Ecology of the Delhi Ridge: Beyond Pristine Glory

Monkeys, of course, are not the only animals living in the Ridge, nor are Mexican mesquite trees the only foliage. On the human side of things, Gujjars and Ods are hardly the only communities residing in or near the Ridge. Still, these four very different groups exemplify the dizzying complexities of Delhi's hills and their ecological history.

It's no coincidence that all these groups are migrants, part of an unceasing flow of people, animals and plants that have shaped Delhi from the very beginning of its history. The Gujjars have made their mark as early rulers, then as pastoralists supplying much-needed milk and meat to the city, then as rebels fighting British rule, then as contractors ensuring a steady supply of stone and sand for the city's construction industry; and now as real estate brokers jumping headlong into the city's thriving casino of land speculation and as politicians brokering formal and informal claims to the city. The Ods, though a much more recent arrival, have still made their mark on the city, first with their innovative digging techniques, which not only fed the construction industry but also laid the groundwork for stadiums, roads, housing complexes and internet connectivity, and then with their tireless fight against displacement.

Vilayati kikar has shaped the city too, though not in unambiguously positive ways. In much of the city, and certainly on the Ridge, it provides a much-needed green cover. It provides firewood as well, not an insignificant consideration given the huge percentage of the population that does not own stoves. But the tree's success has come at ecological costs that will continue to haunt the city, as it wards off other plants and spreads its monotonous monoculture.

And the monkeys. Unexpectedly, they turn out to be an urban species; they serve as a fitting symbol for the Ridge as a uniquely urban forest. But they seem to reflect all the dark sides of human nature. The callousness of the state and its relocation schemes, along with the insatiable, speculation-fueled expansion of real estate, have only aggravated the surly combativeness of the monkeys (and, one could argue, of the humans as well).

But the analogy should not be taken too far. Humans are not just aggressive; not just militaristic; not just violent. The ecological and political history of the Ridge is haunted by all these darknesses, but it offers other lessons as well: the beautifully maintained sacred groves of the Gujjars, of which Mangarbani is the sole remnant; the strong traditions of common ownership and symbiotic customs in grazing zones; the creativity, resilience and political astuteness of the Ods. Even the monkeys aren't all bad; their curiosity remains intact, and they refuse to be jailed, despite the government's many attempts.

The microcosm of Bhagirath Nagar, in which Gujjars and Ods, monkeys and mesquites, all come together, is a far cry from the pristine glory invoked by the Supreme Court. It also bears little resemblance to the Disneyland of the Ridge's rock climbers and mountain bikers. But in all its despair and all its hope, in all its migration and mixing and metamorphoses, it offers a much fuller picture of the Ridge and its shifting ecology.

3 STATE
Warfare, Pageantry, Politics

On the heights of the Northern Ridge sits a forlorn stone pillar. Cars zoom past; ambulances scream through on the way to the nearby Hindu Rao Hospital. In a city with a surfeit of monuments and tourist sites, this one is relatively neglected, though it was once on the main tourist circuit. It still gets the occasional visitor—foreign tourists wandering off the Lonely Planet itinerary, or well-heeled Delhi-ites up early on a Sunday morning for a heritage walk. Like many of the lesser tourist sites in Delhi, it is perpetually locked, a rusty chain encircling its low entrance gate. The pillar itself is cracked and faded, propped up on a rough stone pedestal.

But despite its modest appearance, the pillar has quite a history. It is one of Ashoka's pillars, erected by the great ruler in the third century BCE at the fringes of his Empire to spread word of his reign and his righteousness. But unlike the rock edict in the southern part of the Ridge, which we visited in Chapter 1, North Delhi's Ashokan pillar did not originate in the city; it is not carved from the Ridge's quartzite, though it occupies a prominent place on the Ridge. It is yet another migrant to Delhi.

The pillar's birthplace is the city of Meerut, where it stood happily for over a millennium. In the late fourteenth century, however, another emperor, Firoz Shah Tughlaq, took a liking to the pillar and decided to transport it to Delhi, along with another Ashokan pillar from the town of Ambala. Tughlaq had recently built a new capital city in Delhi, commonly called Firoz Shah Kotla, on the banks of the Yamuna. The Ambala pillar went to the center of his new city, and the Meerut pillar went to the Northern Ridge, near the hunting palace he had built as a getaway from the rigors of courtly life.

The transportation of such weighty stone cylinders was no easy task, especially in those times. But Tughlaq was intent on getting them to Delhi. Ziya al-Din Barani, a chronicler of Tughlaq's reign, describes the immense effort required to dislodge an Ashokan pillar and bring it down to the ground:

> Orders were issued commanding the attendance of all people dwelling in the neighborhood. They were ordered to bring all implements and materials suitable for the work. Directions were issued for bringing parcels of the silk cotton tree. Quantities of this silk cotton were placed round the column, and when the earth at its base was removed, it fell gently over on the bed prepared for it.

This was just the beginning of the onerous task:

> The pillar was then encased from top to bottom in reeds and raw skins so that no damage might accrue to it. A carriage with forty-two wheels was constructed, and ropes were attached to each wheel. Thousands of men hauled at every rope, and after great labor and difficulty the pillar was raised on to the carriage. A strong rope was fastened to each wheel, and 200 men pulled at each of these ropes. By the simultaneous exertions of so many thousand men the carriage was moved.

The pillar arrived in Delhi amid great pomp and celebration, with "the most skillful architects and workmen" employed to fix the pillar in its new home.[1]

This was not the only dislocation the Meerut pillar would suffer.[2] Sometime in the 1710s, during Mughal rule, an accident in a nearby military storehouse caused an enormous explosion. The impact of the blast broke the pillar into five parts, and they lay scattered on the ground for many years.

In 1830, almost 30 years into British rule, the land surrounding the pillar was bought by the Governor-General's Agent in Delhi, William Fraser, in one of the questionable land deals that snatched common land from the hands of Gujjar villagers. Fraser used the land to build an enormous house, with the columns and verandas typical of colonial architecture. It seems he took little note of the stone remnants littering his backyard.

The Gujjars bore a grudge against Fraser, and he was widely disliked by peasants because of the exorbitant taxes he levied on agricultural land. He had more powerful enemies as well. Asked to judge an inheritance dispute between royal brothers, Fraser decided against Shamsuddin Khan, a Mughal nobleman and ruler of a minor principality in present-day Haryana. Shamsuddin, the son of Fraser's now-deceased friend and business partner Ahmad Bakhsh Khan, had looked up to Fraser. But when Shamsuddin went to Fraser's mansion to plead his case, Fraser threw him out. Unable to bear the insult, Shamsuddin hired the sharpshooter Karim Khan to assassinate Fraser. Khan did so as Fraser was returning home from a party on a dark October night in 1835. The assassin and his noble employer were soon caught and summarily executed.

After Fraser's death, a Maratha nobleman named Hindu Rao purchased the house. Hindu Rao, an acquaintance of Fraser's, snapped up the blood-soaked property when it came on the market. Though he originally used the palatial estate as a home for his pet cheetahs, he eventually made the house his primary residence. Shortly after moving in, he found the scattered remains of the Ashokan pillar and quickly recognized their significance. He wrote to the Asiatic Society, a British research organization, and offered the pillar to them as a gift.

The Society, though, was based in Calcutta, then the capital of the British Raj, so the pillar had to undertake yet another journey, one that took it well beyond its original home of Meerut. However, the officials in charge of transportation, more pragmatic than Firoz Shah Tughlaq, decided that it would be too costly to ship all five pieces. Instead, they only shipped the piece that had the Ashokan inscription on it.

In 1866, for reasons unknown, the British shipped the pillar fragment back to Delhi, and in the following year, British engineers put all the pieces back together again and placed it on the stone pedestal where it now stands. Never known for their modesty, the Raj rulers placed a sign at the base of the pillar, announcing their role in putting it right again.

After Independence, the postcolonial Archaeological Survey of India (ASI) put their own stamp on the monument, in their typically minimalist way. They erected the fence that now runs around the perimeter of the pillar, as well as the signature red-and-blue ASI sign, which declares that the site is protected by law, but maddeningly tells nothing about the monument itself.

The pillar may appear mute now, but its history speaks volumes about the way the state has articulated its power in Delhi, or rather, how multiple competing states have struggled to assert their power. The history of governance in Delhi is hardly continuous, after all, and hardly uncontested. In the archaeological record, these contestations first become visible with the fortresses and monuments of Mehrauli, which rose in the early days of Arid Zone dynamism. As we have seen, the choices of these early states were partially determined by ecology, both of the Arid Zone in general and of the Ridge in particular. But once Delhi became established as a capital, political processes acquired their own momentum, as the prestige of previous rulers was appropriated by their successors.

Traditional histories speak of seven cities arising in Delhi. They followed the same pattern: a new set of rulers would vanquish the old guard and build themselves a new capital in Delhi, only to be vanquished by an even newer set of rulers. Each new state would assert its superiority to the old one, while simultaneously appropriating old symbols of power and glory. Hence the profusion, and the relativity, of sites described as "old" in Delhi. Currently, Shahjahanabad is known as "Old Delhi", but for the eighteenth-century inhabitants of Shahjahanabad, "Old Delhi" meant the crumbling ramparts erected by Firoz Shah Tughlaq, whose city was new in comparison to his grandfather's Tughlaqabad, and on and on.

Sometimes a new settlement would engulf an old one in quite literal ways. When the Chauhans defeated their fellow Rajputs, the Tomars, to take control of Delhi in the twelfth century, the new rulers simply expanded the walls of the old Lal Kot to build Qila Rai Pithora. Mohammad bin Tughlaq took this logic to its extreme, seeking to enclose four older settlements of Delhi (including Lal Kot and Qila Rai Pithora) into an enormous new walled city.

Some rulers, on the other hand, wanted to make a fresh start. Ghiyasuddin Tughlaq, founder of the Tughlaq dynasty and Mohammad bin Tughlaq's father, moved to the southeastern corner of the Ridge for his new citadel of Tughlaqabad. His grandson Firoz Shah abandoned the Ridge altogether for *his* new city, although he then chose the northern portion of the Ridge, then completely rural, for the location of his hunting palace. But whether the new and the old were physically overlapping or far-flung, the accumulation of cities merged to form a conceptual whole: Delhi as the seat of power, Delhi as the focal point of the state.

Each new state has presided over violence and has come to power in the face of uncertainty and instability. To establish its sovereignty, the new state has downplayed this uncertainty, papering it over with pronouncements and displays of power and legitimacy. Often, this has taken the form of invoking the past glory of former rulers, repurposing and re-appropriating their symbols of power. As the Ashokan pillar suggests, the Ridge, with its commanding heights, was often the site of this.

The pillar is a testament to both the strength and the shakiness of state power in Delhi. Having been dislodged from its ancient resting place and relocated to attest to Firoz Shah Tughlaq's medieval power, the pillar was, as we've seen, ignored and accidentally dismembered by the Mughals, recognized by a lone Maratha nobleman, handed over to the British, and reclaimed by the postcolonial government, which has now banished it to heritage purgatory: not the complete disregard shown to some of Delhi's crumbling historical sites, but not the care lavished on revenue-generating tourist sites like Qutb Minar either.

Despite their sometimes-lackadaisical approach to the relics of Delhi's previous incarnations, the leaders of independent India have clearly recognized the benefits of invoking the might and glory of past rulers. It's no coincidence that the two chief occasions of government pomp (Independence Day and Republic Day) are celebrated in front of the massive monuments to power built by the Mughals and the British, respectively.

And while specific pieces of built heritage, like the Ashokan pillar, are relatively neglected, sites of "natural heritage", like the Ridge, are increasingly (if mistakenly) seen by the state as markers of "pristine glory" that must be preserved. This is not simply an ecological argument; various government officials, starting with the British and continuing to the present day, have explicitly linked the majesty of an afforested Ridge with the impressiveness and legitimacy of Delhi as a capital city.

The state's entanglement with the Ridge has been a complex one. Elements of this entanglement have already emerged in previous chapters: the state's thirst for stones and minerals to be used for construction; the state's need to establish a threshold between pastoral services and agricultural surplus; even the state's desperation to relocate monkeys. But this hardly scratches the surface of an ever-changing, millennia-old relationship between various state powers and the set of hillocks and plateaus known as the Delhi Ridge.

This relationship is especially challenging to trace because "the state" is both historically variable and notoriously difficult to define. The state established by the Tomars in medieval Delhi had vastly different functions, ideologies and powers than the one currently governing from Raisina Hill. Premodern states, though they certainly had their complexities, were focused on a relatively limited set of goals, with particular priority given to raising armies, collecting taxes, and suppressing rebellions. Modern states have certainly not abandoned these goals, but they are far more ambitious. Their interventions are more sweeping, in ways both good (public health, social security, universal education) and bad (surveillance, censorship, slum demolitions).

Struggles over state power in Delhi—over its exercise and over its representation—are not just abstract debates rehearsed in history books, academic journals and political tracts. They have played out in physical spaces as well, whether in the reconstruction of an old broken pillar or in the assassination of a powerful political figure. State power has been made and unmade, projected and rejected, on the Ridge.

Given the multifarious role of the state in Delhi's long history, this chapter has been divided into three parts, each reflecting a major function of the state over the centuries: military conquest; building and planning; and the provision of public services.

PART I: SOLDIERS

Blood and Glory

Before the state can perform its functions, it must first come into power. In Delhi, at any rate before 1947, this has meant military conquest. Armed power is not just necessary in the beginning, to establish power, it is an integral part of any state's rule.[3] Whatever else a state does, whatever its broader goals and ambitions may be, it is nothing without the power to enforce its rule.

The role of armed violence in establishing state power goes back deep into Delhi's mythical past. The Khandava bloodbath, described in a previous chapter, represents, in one interpretation, at least, the military victory of the invading Pandavas over the native forest-dwellers of the Delhi area. Violence was thus a key precondition for the founding of Indraprastha. But soon after this slaughter, the Pandavas were sent into exile, with little opportunity to enjoy their new city. In this, too, they serve as a mythical prototype for state power in Delhi, which has been notoriously difficult to hold.

There's even an old apocryphal saying: "He who builds a new city in Delhi will soon lose it." The rapid succession of kings and emperors suggests that few states have been as stable and solid as their propaganda boasts. Especially because of its strong historical associations with empire and might, Delhi has been a coveted possession for rulers wishing to make their mark. And because of the wealth it amassed as a capital city, it was a favorite target of invading armies. Of course, invaders rarely considered themselves invaders; they saw themselves as noble conquerors and statesmen fighting for a righteous cause— from the mythical Pandavas to the Rajputs, Sultans, Mughals and British. And if they were really successful, they could convince the public (and the historians) to consider them as such.

The Ridge has thus been both a refuge against "invaders" and a staging ground for them. In the battles of Delhi, the heights of the Ridge have played a dual role: in a very literal way, the physical heights of the Ridge have aided armies by giving them a vantage point from which to survey and attack the enemy camp. More metaphorically, the heights have been used to shout out the glory of new regimes, or old ones struggling to maintain their legitimacy.

Medieval Militaries

When Delhi first emerged as a medieval city, the Southern Ridge was attractive precisely because of its commanding heights. (Technically, in present-day terminology, this is the South-Central Ridge, but I'll use the term "Southern Ridge" here for simplicity's sake.) A Ridge-top fortress, however, was no guarantee of staying power, as the Tomars and Chauhans quickly found out. When Mohammad Ghori, Sultan of the Ghurid Empire, defeated Prithviraj Chauhan in 1192, he kept the Rajput's Delhi outpost intact, and his successors expanded from this hilltop base. Chroniclers of Delhi's history—from Ghori's contemporaries to British colonial historians to present-day Indian scholars—have generally portrayed this as a decisive changing of the guard, a sudden and definitive entrance of a strong new force in Indian politics; whether this was seen as a calamity or a triumph depended on the observer.

But the consolidation of Delhi as a new power center was hardly inevitable. Mohammad Ghori was caught in a web of rivalries and intrigue, centered around his family's base in Ghor, a city in what is now Afghanistan. (It is worth noting in passing that this family, too, came from pastoral roots and was initially derided as uncultured and uncouth.) Ghori, after a decade leading his army to a string of military victories in Punjab, Gujarat, Bihar and Bengal (1186–96), abruptly withdrew from the subcontinent, leaving it in the hands of various commanders from his military retinue.

One of Ghori's slaves, Qutbuddin Aibak, eventually took charge of Delhi; this was the beginning of the *Mamluk* or Slave Dynasty in India. Other slaves were left in charge of other parts of the subcontinent. But these were no ordinary slaves. Certainly, they do not fit the image of grinding, lifelong oppression and exploitation that the term "slavery" usually evokes. As slaves, they were bound to their master, but they were also given the opportunity to rise to positions of power, after which they could be freed.

Though their combination of low social status and sometimes-immense political power seems paradoxical, it had a clear logic to it. Rulers like Ghori bought slaves like Qutbuddin Aibak when they were young, and indoctrinated them in everything from court etiquette to martial arts to religious practice. Ghori, like many of his contemporaries, was particularly interested in amassing Turkish slaves, who were reputedly fearsome fighters and loyal servants. As a poem from that era puts: "One obedient slave is better than 300 sons, for the latter desire their father's death, the former his master's glory."[4]

This was the logic that led Mohammad Ghori to install the Turkish slave Qutbuddin Aibak as his Governor in Delhi. But the slaves, who had already proved their military worth and their shrewd ability to navigate the politics of conquest, were hardly docile. While all still pledged their loyalty to Ghori, they began to jockey for position and consolidate power in their fiefdoms. Aibak was especially aggressive, and his power plays were particularly significant for Delhi. In 1192, Ghori had left Delhi nominally in charge of a vassal from the Chauhan family, but Aibak seized the Ridge-top outpost from the vassal before the year had ended. There are stray references to Ghori's dissatisfaction with Aibak's impertinence, but, through visits to Ghazni and other placatory gestures, the slave was able to stay in his master's confidence.

After Ghori's death in 1206, Aibak positioned himself as the rightful Sultan of Delhi, carrying on the rule of the Ghurid Empire, despite the presence of several competitors. Unlike the Chauhan rulers he had displaced, he promoted Delhi as a capital city. His Ridge-top fortress, inherited from the Rajputs, was projected as the center of an important dynasty. Court chroniclers, writing in the royal language of Persian, backed up this claim. The courtier Fakhr-i Muddabir dutifully records that Aibak is the "hero and world conqueror of Hind" and that Ghori named him "as his deputy and heir in the capital of Hindustan".[5] Thus Aibak, through his scribes, initiated a long tradition of invoking the glory and power of Delhi.

However, Aibak was hardly the unchallenged ruler of northern India, no matter how much he proclaimed himself to be. And he died just four years after Mohammad Ghori, so he had little chance to consolidate his power and challenge his rivals across northern India. By the time of his death, Aibak had amassed his own retinue of slaves, and his death unleashed a fierce power struggle between them. A powerful Turkish slave named Shamsuddin Iltutmish finally triumphed over his rivals.

Though he had vanquished his internal foes, Iltutmish found himself surrounded by a sea of external enemies. The Delhi Sultanate was arguably the weakest of four powers that had emerged in the aftermath of Mohammad Ghori's north Indian conquests. While Iltutmish was fighting his rivals, these other powers had eaten considerably into the territorial holdings of the Delhi Sultanate. Iltutmish spent the first 20 years of his reign clawing back these territories and counterattacking to expand his territories. By 1229, he had defeated all his major rivals, and for the first time, the Sultan of Delhi, from his seat on the Southern Ridge, could reasonably proclaim that his was the strongest political force in North India.

This claim was backed by new migrants to the city, who had their own reasons for promoting Delhi's centrality. For Iltutmish's rule was marked not just by strife between former allies, but also by larger calamities. From 1219 onwards, the Muslim elite in Transoxania and Afghanistan were driven from their lands by Mongol invasions. These elites suffered not just the physical trauma and terror of Mongol conquest and violence, but also the indignity of being driven away by forces they saw as barbaric, idolatrous and boorishly uncultured.

Iltutmish, seeking ideological support to complement his military strength, welcomed the noblemen, traders and religious scholars who had fled Mongol destruction, offering them land grants and government posts. Atop the Southern Ridge, he attempted to build a new center of Muslim culture, piety and refinement. As one of the emigres put it, Iltutmish had transformed Delhi into *Qubbat al-Islam*: a sanctuary for Islam. But *Qubbat* means more than just sanctuary; it suggests an axis, or a center. If Aibak claimed that Delhi was the center of Sultanate power in Hindustan, Iltutmish went a step further and promoted the city as a focal point for an entire faith.[6]

But all was not well, even at the height of Iltutmish's reign. Military threats to his regime never entirely disappeared, though they were increasing confined to the edges of his empire. Closer home there were rumblings of discontent from the hermitages of Sufi saints. These mystical practitioners of Islam had little patience for the orthodoxies of the city's religious scholars. Chief among the Sufis was the renowned mystic Qutbuddin Bakhtiar Kaki, who also fled from the Mongols and settled in Delhi after a brief stint in Baghdad. His fame would eventually rival that of the Sultans, and he chafed under the control of newly powerful clerics.

This discontent found dramatic expression at the tail end of Iltutmish's reign. In 1236, a Sufi saint named Nur Turk, who had a sizable following of devotees from Gujarat, Sindh and Delhi, led an attack on the city's main congregational mosque. Apologists for the regime claimed that Nur Turk was simply a heretic, bent on destroying a pious Muslim community because of his rebellious ways. But commenting on this event a generation later, the famed Sufi saint Nizamuddin Auliya (whose spiritual master, Baba Farid, was a follower of Qutbuddin Bakhtiar Kaki) absolved Nur Turk of any heresy.[7] Instead, he argued that the mosque attack was a reaction to the indiscretions of the orthodox religious scholars, who had been swayed by the material riches of Delhi.

Nur Turk's uprising was quickly quelled, but it serves as a potent reminder of the vicissitudes of power in the Delhi Sultanate, and the constant threats it faced both from within and without. On the whole, the Delhi Sultanate was not a stable empire but rather a series of dynasties whose power fluctuated dramatically: after the Slave Dynasty, Delhi was ruled by the Khiljis, then the Tughlaqs, then the Sayyids and the Lodis. The last two dynasties were particularly weak. Later accounts, such as the ones compiled by the Mughals, who ushered in a period of relative stability, consider Firoz Shah Tughlaq to be the Delhi Sultanate's last leader of any note.

But these same Mughal accounts also stress Delhi's "axial importance in subcontinental politics".[8] Despite the rapidly-diminishing power of the Sultanate in its later years, the propaganda efforts started by Qutbuddin Aibak on the Southern Ridge centuries earlier proved effective; Delhi was regarded, even by skeptical outsiders, as a strong political capital, indeed, as it still is.

Meanwhile, on the Northern Ridge

The Southern Ridge saw the first articulations of Delhi as a powerful political capital and as a refuge for a threatened civilization. The Northern Ridge came to prominence later, but its history reveals similar patterns: military struggle and considerable violence, followed by attempts to justify and legitimize state control, alongside more subterranean expressions of local tradition and resistance.

The main event in this history is the gruesome bloodshed of 1857, but this colonial violence has an uncanny medieval echo. In December 1398, the feared warrior Timur descended on Delhi with his army. Timur was yet another conqueror with nomadic Central Asian roots; he combined Mongol military prowess with Turkic and Persian taxation systems. He presented himself as a restorer of the Mongol Empire and an heir (in spirit, if not in biology) to Genghis Khan. This was hardly reassuring to the Sultans of Delhi, who had long seen themselves as a refuge from the predations of the Mongols.

Timur and his troops swept through northern India with ease, taking advantage of the disintegrating Delhi Sultanate. Just as Mohammad Ghori and Qutbuddin Aibak had benefited from intra-Rajput rivalries, Timur profited from intra-Sultanate discord. He arrived in India only ten years after Firoz Shah Tughlaq died, but Delhi had since witnessed the quick succession of six Sultans, with the Sultanate splitting in 1394, leaving the new ruler of Delhi, Mahmud Shah Tughlaq, fatally weakened.

When Timur arrived in Delhi, he sent an advance guard of troops to inspect Firoz Shah Tughlaq's hunting lodge on the Northern Ridge, which Timur described as a "fine building". But his appreciation for architecture gave way to chilling orders for his troops: "to plunder and destroy, and to kill every one they met."[9] Timur later joined the advance guard to view the palace himself and to use the vantage point of the Ridge for reconnaissance.

Later, he returned to the foot of the Northern Ridge with his entire army, and they set up a heavily fortified camp as a springboard for their attack on Delhi. Though this campsite is now in the midst of Delhi, at that time it was deep in its hinterland, roughly five miles north of the urban enclaves that the Tughlaqs had built, which were themselves located just north of the old Slave Dynasty fortresses on the Southern Ridge. Timur's plan was to draw out Mahmud Shah's army and to defeat them on the plain between the Northern Ridge and the city of Delhi. His troops had warned him of the massive, fully armored war elephants that formed the core of the Sultanate defenses.

In response, Timur ordered that all the trees in the area be cut down (yet more evidence of centuries-old deforestation on the Ridge) and used to reinforce the trenches that his troops had dug around the camp. Timur also instructed his troops to tie water buffaloes to the newly erected fences as a further line of defense against the elephants.

On the morning of the decisive battle, Timur mounted his horse and rode up to the top of the Northern Ridge, to survey the enemy army as it approached. Satisfied with what he saw, he rode down and led his troops into battle, driving the Sultan's army all the way back to the gates of the city, where they conceded defeat.

The battle was just the beginning of Delhi's suffering. Timur set up his throne in the Idgah, located on the far outskirts of the Tughlaq city atop the Southern Ridge. There, he received the "learned Muslims" of the city, who were given "the honor" of kissing his throne and who begged Timur to spare their city. He amused himself by surveying the elephants that the Sultan had left behind; the elephants' drivers, eager to placate Timur, had the elephants bow their heads as Timur passed. The conqueror, pleased, had the elephants sent to his various strongholds throughout Central Asia, with five of the best ones dispatched to his capital city, Samarkand. Timur also threw lavish parties for his triumphant army, bringing in Turkish and Arab musicians and dancers, and serving "wine, sherbet, sweetmeats, and all kinds of bread and meat".[10]

But there was the serious business of plunder to carry out as well. Timur sent his men around the city to collect money, as well as grains, oil, sugar and flour. The people of Delhi resisted, no doubt alarmed by the presence of marauding troops who, even by Timur's admission, were laying "riotous hands" on the city and its riches. The resistance was short-lived. Timur's troops massacred the population, burned down much of the town and plundered whatever they could.

In the fighting, a group of rebels escaped from the mayhem and gathered on the Southern Ridge, venturing beyond the Idgah to take refuge in the Slave Dynasty-era congregational mosque, which had earlier weathered Nur Turk's attack and was, at that point, considered part of "Old Delhi". The rebels started to stockpile weapons and food supplies, but Timur found out about their plot and sent his troops, who "attacked these infidels and put them to death, after which Old Delhi was plundered".

Justifying this gruesome multi-day massacre, Timur lapses into the passive tense, merely noting that "certain incidents occurred" and the "flames of strife" had been lit. He then falls back on a religious justification:

It was therefore my earnest wish that no evil might happen to the people of the place. But it was ordained by God that the city should be ruined. He therefore inspired the infidel inhabitants with a spirit of resistance, so that they brought on themselves that fate which was inevitable.[11]

Many modern historians, and not just Hindu nationalist ones, have accepted Timur at his word, portraying him as yet one more sword-wielding Muslim fanatic.[12] Yet whatever his inner religious beliefs (and there is ample evidence that he wore his religion lightly, parading it about at times and dropping it when convenient),[13] it is clear that Timur's military conquests were serving larger political and economic prerogatives.

Timur's raids and conquests in India were not just the result of random cruelty (though they were certainly cruel). They were part of the logic of empire, which centered on Timur's capital of Samarkand. It's no coincidence that he sent Delhi's best elephants there. And it was not just elephants. Timur also records that he rounded up all of Delhi's "artisans and clever mechanics who were masters of their respective crafts"[14] and sent them to different cities of his empire, so that they could aid their economic growth. For himself, he took all the builders and stonemasons of the city and sent them to Samarkand, where he was building his own congregational mosque to memorialize his rule. The historian Satish Chandra explains that this process applied not just to Delhi but to Timur's other conquests in Baghdad, Damascus and beyond, and that Timur's interests extended to culture as well as trade. In this way, he amassed a group of skilled painters, calligraphers, musicians, historians, architects, silk-weavers, bow-makers, masons, metalworkers, gem-cutters and more.

Chandra continues: "Timur was not merely a Central Asian, semi-nomadic leader intent on carving out as large an empire as possible. His conquests were aimed at...unifying the principle Asian trade routes, overland and overseas, under his control."[15] From this perspective, Timur's sacking of Delhi and his other exploits in northern India largely served as a way to safeguard key trade overland routes connecting Central Asia to India, as well as to secure ports on the western Indian coast. An Italian traveler of that era, Nicolo Conti, notes that travel was safe between Egypt and India because the entire zone was controlled by Timur.[16]

For Central Asian traders, and for the occasional European adventurer, Timur's reign thus held out the prospect of stability, prosperity and thriving economic and social exchanges. For the inhabitants of Delhi, whether Muslim or non-Muslim, his reign was an unmitigated disaster, bringing destruction, bloodshed and chaos. Timur's beautiful mosques, built in part by Delhi's artisans, still stand in present-day Uzbekistan, where the fearsome ruler is still remembered fondly. (When the Soviet Union fell and Uzbekistan became independent, the new government replaced a statue of Karl Marx with one of Timur.) These mosques serve as a testament to Walter Benjamin's famous dictum, "There is no document of culture which is not at the same time a document of barbarism."

For Timur, the Ridge was just a temporary resting ground, a place to survey the battlefield and inspect the architectural glories of the dynasty he was about to destroy. For the early Sultans of Delhi, on the other hand, the Ridge was their sanctuary, the axis of their faith and their state. Timur's brutal raid marked a low point for the Delhi Sultanate, from which it never truly recovered.

The Mughals, who succeeded the Sultans, remained attached to the Ridge, and specifically to the Southern Ridge that hosted Delhi's oldest fortresses. In Mughal times, the focal point of "Old Delhi" was the shrine to the Sufi saint whose influence Iltutmish had once tried to curtail: Qutbuddin Bakhtiar Kaki, affectionately known as Qutb Sahib. Qutb Sahib's repute had grown posthumously, largely because of the fame of his disciple's disciple, Nizamuddin. By the time Babur came to India, Qutb Sahib's shrine had become a major pilgrimage point, with a bustling local community dedicated to the glory of the saint. In the centuries that followed, several Mughal emperors chose to be buried close to the Ridge-top shrine.

The Ridge, then, has borne witness to shifting imperial dynamics. As increasingly powerful empires have set their sights on Delhi, their rulers have not forgotten the material and symbolic importance of the Ridge. It has been a crucial node in economic, military and religious circuits that have spanned increasingly large parts of the world.

A New Empire, A New Army

Even for the massive British Empire, which, at its peak, engulfed much of the world and towered above all other powers, the small piece of land known as the Delhi Ridge played an outsize role. For patriotic Britons in the late nineteenth century, "The Ridge" was synonymous with sacrifice, heroics, bloodshed and redemption, a potent symbol of the trials and the triumphs of empire.

This was an improbable development, especially since Delhi was a relatively late addition to British India. The British did not enter India by land. In contrast to their imperial predecessors (the Delhi Sultans, the Timurids, the Mughals), the British did not come upon Delhi as a crucial threshold for their initial Indian conquests. In keeping with its origins as a trading firm, the British East India Company first engaged with India via its coastal ports. As it morphed from a business-oriented company to an unwieldy state power, the East India Company organized itself around three port cities—Calcutta (now Kolkata), Madras (now Chennai), and Bombay (now Mumbai)—with ever-growing Presidencies spreading out from those centers. In its expansionary efforts, it was backed by its own private armies, largely composed of Indian mercenaries, though with a sizable number of Irish and English recruits as well. Just as the three Presidencies were relatively autonomous, so too were the three armies raised to defend them and expand their borders; the biggest of these armies, and the most important, was the Bengal Army, which had its roots in the East India Company's most important urban center: Calcutta.

When General Gerard Lake defeated the Marathas, who had been hounding the Mughals, and marched into Delhi as the defender-turned-usurper of the Mughal Empire, he did so as the head of the

Bengal Army of the East India Company. The British recognized that Delhi had long been an imperial capital, but they also knew that the Mughals had been drained of all their power, and they thought of the rituals and procedures of courtly life in Delhi as mere relics of a spent empire. They thus allowed the Mughal emperors (now emperors in name only) to retain some vestige of their symbolic prestige, while eviscerating whatever dwindling power they still had.

In 1828, in keeping with this mindset, the British humored the Mughal Emperor, Akbar Shah II, and moved their army's European troops outside of the walled Mughal city of Shahjahanabad, and into a cantonment on the western slopes of the Northern Ridge. Most of the cantonment's buildings were at the base of the Ridge, but a few were atop the Ridge, including Flagstaff Tower, located on the highest point on the Northern Ridge. This structure was used as a lookout tower, with the British flag flying proudly atop it.

The British felt justified in making this move since the Mughals had no military power; British officials were more worried about attacks from the Sikh Empire, against whom they were soon to fight two wars. Their new position on the Northern Ridge was ideal for fending off attacks from the northwest, where the Sikh armies would likely amass. To further fortify their position, the British strengthened the walls encircling Shahjahanabad, reinforcing with stone and brick the mud structures that the Mughals had built.

The British were thus completely taken by surprise when the biggest threat to their rule came from inside the walled city itself. On 11 May 1857, a group of rebel soldiers forced their way into Shahjahanabad and set up base in the city before anyone could stop them. The troops had come from Meerut, where, the previous day, they had mutinied and killed their commanding officers, along with both Indian and European civilians living in the area.

The arrival of the Meerut troops in Delhi was perhaps the defining moment of the 1857 Uprising. Although the rebellion had many causes (including the discontent felt by groups like the Gujjars) and took different paths in different parts of the country, Delhi was its symbolic center. This, at least, can be inferred from the actions of the Meerut troops, as well as many other rebels from across North

India who followed in their footsteps and converged in Delhi. In their uprising against the British, the Meerut troops sought legitimacy from the very symbols that the British had thought were hopelessly spent: the prestige of the Mughals and the imperial lineage of Delhi. From Delhi, the rebels found a symbolic center for what was to become the biggest uprising against the most powerful empire in the nineteenth-century world.[17]

The soldiers from Meerut came to Delhi to get the blessing of the Mughal emperor, Bahadur Shah Zafar, in their rebellion against the British. Confronted with the unruly troops, a reluctant Zafar had little choice but to agree. The uprising came as a rude shock to the British, and it accelerated a hardening of British attitudes towards the people they sought to rule in India. The Indian subject was increasingly seen as irremediably savage, and the violence of the early days of the Uprising was invoked as proof of this.

A chief symbol of supposed Indian savagery was the Flagstaff Tower on the Northern Ridge. By the afternoon of 11 May 1857, the rebelling armies had swept through Shahjahanabad and had ransacked the cantonment outside the city walls. Flagstaff Tower became a place of refuge for the British families fleeing both the walled city and the cantonment. Women and children were packed into the sweltering interior of the tower, and several passed out due to the heat and stress.

A handful of British army officials guarded the tower, and they argued over the best course of action: should they stay in the tower until nightfall, or should they retreat immediately? As they debated this, a bullock cart made its way up the Ridge to the tower, carrying in it the mutilated bodies of British soldiers who had been killed in the walled city. Although the cart had been sent by another British officer, and was on its way to the cantonment, the understandably jittery families in the tower assumed it was an act of psychological warfare on the part of the rebels. This ended any argument: the families poured out of the tower and began their slow, tortuous escape.

It took the British almost a month to reclaim the Ridge. On 8 June, under Major-General Henry Barnard, the Delhi Field Force entered the Delhi region, decisively defeating the rebel troops at a western outpost of the city. From there, the British troops moved to the Ridge, shooting everyone in their path, including a young fakir who had attempted to take refuge in an old ruined mosque that had been built in the time of Firoz Shah Tughlaq. At Flagstaff Tower, they faced fierce resistance, but eventually overpowered the rebels, who retreated to the safety of Shahjahanabad. The bullock cart full of corpses still sat beside the tower; in the scorching summer heat, the bodies had become mere skeletons. In this grim setting, the British ran their flag up the tower's flagpole and considered their options.

Although some trigger-happy British officials, both at the time and in later accounts, faulted Barnard for stopping at the Ridge and not following the rebels into the city, he clearly made the prudent choice. The rebel troops far outnumbered his own, and they were protected by the strong city walls that the British themselves had reinforced. The Ridge was a key strategic site; it overlooked Shahjahanabad and allowed the British troops to survey rebel movements and fight off any attacks from a position of strength.

But the Ridge had its drawbacks as well. In the peak of the summer, it was blazing hot and almost entirely barren, with only the occasional stunted tree. There was no reliable water source nearby, and the nearest available water, coming from the Yamuna canal, had a horrible, putrid taste. Several of the troops died of heatstroke. And then there were the military dangers: the British encampments on the Ridge were clearly visible from the walled city, and the rebels soon began to shell the British strongholds on the Ridge with alarming accuracy. One of these strongholds was the mansion built by William Fraser, who had been assassinated 22 years earlier. By now, the house was known by the name of its subsequent owner; in British correspondences, it is invariably referred to as Hindoo Rao's house.

The physical features of the Ridge had psychological impacts as well. From their various lookouts on the Ridge, the British troops could see new sets of rebel troops streaming into the walled city nearly every day. This sight was hardly reassuring to the British, especially as it underscored their own precarious position. Although they initially hoped for quick reinforcements so that they could lay siege to Delhi, it soon became evident that supporting troops were still a long way away. Many of the troops who would eventually ensure the British victory had not even been gathered and mobilized. Given the scope of the rebellion, British officers were casting their nets far and wide to gather the troops and the equipment they needed to quell it. The soldiers on the Ridge were, for the moment, on their own.

Because of their favorable position on the heights of the Ridge, the British were able to repel the waves of attacks launched by the rebels. However, as the days on the Ridge turned into weeks and then into months, the British troops faced a slow transition from military threats to epidemiological ones. Their very success in cutting down rebel troops led to another grim problem; decomposing bodies littered the side of the Ridge, attracting swarms of disease-carrying flies and emitting an intolerable stench. To make matters worse, the monsoon, which had arrived on 27 June, turned many spots on the Ridge into fetid swamplands, unleashing plagues of snakes and scorpions. On 5 July, Barnard died of cholera. By the end of July, more soldiers were dying of cholera than of bullet wounds. Their attempt to quell the Uprising had morphed into a war of attrition against more implacable enemies.

In the letters that soldiers sent out from the Ridge, especially to their wives, they maintained a stiff upper lip, minimizing the precariousness of their situation. But hints of the dangers they faced inevitably surface, as in the case of one letter from an officer named Keith Young to his wife:

It has been a very hot day, but a refreshing shower an hour or two ago has made it considerably cooler, and there is now a pleasant breeze blowing, and thunder rumbling in the distance. I wish it would come down a good plump of rain, and it would serve, too, to put out the fire in one of our batteries, which caught fire just now, and seems inclined not to allow itself to be put out.[18]

Relief finally arrived on 14 August, in the form of a mercurial and, by some accounts, psychopathic British officer named John Nicholson. Nicholson had spent the previous three months repressing discontent in Punjab and rounding up troops. Remarkably, given the fact that the British had just fought a war against the Sikh Empire, and would fight another just one year after the Uprising ended, a large percentage of Nicholson's troops were Sikhs. Their long-standing hatred of the Mughals, who had killed several Sikh gurus, seemed to outweigh their relatively new grudge against the British. Nicholson led the advance guard into Delhi; behind him was an enormous "siege train", an eight-mile long convoy of troops, guns and ammunition, with the larger guns pulled by teams of elephants.

On 4 September, the siege train finally arrived on the Ridge. By 12 September, all 60 of the howitzers that had been lugged to the Ridge were firing non-stop on the city walls. And on 14 September, Nicholson led the attack on the city. In the chaos of the fighting, a rebel sniper shot Nicholson in the chest. A young officer arranged Nicholson's transport back to the Ridge, and he spent the last days of his life in the camp's makeshift infirmary, as fighting raged in the city and the outcome of the battle seemed increasingly uncertain.

Archdale Wilson, the new Commander-in-Chief, was unnerved by the losses the British were absorbing. He knew Nicholson had been shot, and he was also aware that one of the rebel leaders was leading a contingent up the Ridge to Hindu Rao's House, which, if successful in its mission, would completely encircle the British camp and cut off contact between the British soldiers in the city and their reinforcements, supplies and medical facilities on the Ridge. Several times, Wilson contemplated a full retreat.

However, after a week of chaotic, bloody street fighting, the British broke the back of rebel resistance. On 20 September, they stormed the Red Fort, the center of Mughal power, and began a systematic massacre of the Delhi population. Within days, a large number of the city's inhabitants, 150,000 people, had either been killed or forced to flee. Huge sections of the city were razed, and the triumphant troops looted and plundered with abandon, following in the footsteps of Timur's troops 450 years earlier.

The British, at least in their official pronouncements, felt few scruples about the violence and trauma they had unleashed. If any-thing, their tone was celebratory, with a grim sense of satisfaction at having shown the native his proper place. Echoing Timur's language, the British focused on the supposed religious barbarity of the rebels, in contrast to their own spiritual uprightness (here Christianity took the place of Timur's Islam). Nowhere was this more evident than on the Ridge. It was the "Mutiny", in fact, that turned the ridge north of Shahjahanabad into "The Ridge", a proper noun, recognized by patriotic souls across the British Empire. Although today "the Ridge" generally refers to all of the Aravalli hills in Delhi, the phrase originally designated the small ridge where the British pitched their tents during the Uprising. Even before the Uprising had ended, tales of British heroism and hardship on the Ridge were becoming the stuff of legend.

Evidence of this consecration could be found in newspapers articles, letters and speeches throughout the British Empire, with especially strident voices coming from London, the center of the empire. If the city stood, in the British mind, for savagery, debauchery and near-apocalyptic evil, then the Ridge, in its austerity, symbolized British righteousness and heroism. This idealization of the Ridge took concrete form in 1863, when British officials constructed a garish Mutiny Memorial to commemorate the troops that had fallen in the fighting. The Memorial, constructed in Gothic style, mimics the form of a church cathedral, with a large cross at the top, in case anyone missed the symbolism.

Yet most of those actually fallen, whom the Memorial ostensibly celebrated, had not entered the fight for the sake of the British flag, let alone for Christianity. Four-fifths of the soldiers on the British side were Indian, with many Sikh and Muslim mercenaries filling the ranks. And even on the British side, there were fissures between the Irish and English and between troops from the East India Company and from the Queen's Army. Some Company troops even scrawled graffiti on the walls of a crumbling ruin on the Ridge condemning a sergeant for watering down their grog.[19]

The Ridge-top scrawlings of restive soldiers are drowned out by the stately pronouncements of officialdom, in the form of the Mutiny Memorial. But even this memorial, built to commemorate one of the most significant moments in British imperial history, eventually faded from popular memory. It now receives precious few visitors and hardly fits into the public imagination of the city. It is one more record of a state's failed attempt to immortalize itself.

Perhaps this is to be expected, more than 70 years after Independence. What is more surprising is the quickness with which the memories of the "Mutiny" were forgotten. As early as 1902, British officials were bemoaning the lack of attention given to the sacred Ridge and urging the public to rediscover it. H. C. Fanshawe, a former Commissioner of Delhi District, made such a plea in his guidebook to Delhi. In many ways, Fanshawe's book, entitled *Delhi: Past and Present*, resembles a contemporary *Lonely Planet*: it suggests three- and five-day itineraries for the time-pressed; dividing the city into different, easily-visited sections, it provides maps and directions, and even puts important site names in bold font. But the underlying spirit of Fanshawe's guide is patriotic, not commercial. He writes:

> *I would venture to hope that the present volume will afford to visitors to Delhi not only a clear guide to all that is to be seen there, but also an intelligent record of the history of the place in all its various phases, and will help to secure a permanent memory of such and of many others, for the great and gallant feats of arms performed before Delhi in the summer of 1857, by a very small force under the most arduous and trying conditions. I cannot but think that the recollection of this feat, not yet fifty years ago, has become somewhat unduly dimmed.*[20]

The centerpiece of his book is a chapter on "Delhi in 1857", an exhaustive account of all the sites, on the Ridge and elsewhere, where fighting took place, as well as the surrounding gravestones, ruins and monuments, along with an exceedingly detailed narration of the events of May through September, interspersed with multi-page excerpts from the journals and notes of various military officials who were part of the Delhi fighting. For Fanshawe, the "Mutiny" was the pinnacle of Delhi history. The preceding centuries, which he hastily describes in the barest of outlines, are mere ornamentation and backdrop for the true story of Delhi's glory.

Delhi continued to be important for the British, especially after it officially became the capital of British India in 1911. But the establishment of a new imperial capital brought the city's center of gravity away from the Northern Ridge and the former sites of the 1857 fighting. Though the planners of New Delhi acknowledged the sanctity of the (Northern) Ridge, their attention was mostly elsewhere.

Ironically, it was the post-Independence government that was to bring the Mutiny Memorial back into prominence, albeit briefly. In August 1972, to celebrate 25 years of Indian independence, the government renamed the monument "Ajitgarh" or, roughly, "Victory Fort". Beside the British inscriptions, the Indian government added its own, a sort of exegetical footnote: "The 'enemy' of the inscriptions of this monument were those who rose against colonial rule and fought bravely for national liberation in 1857." This was government appropriation of the most unimaginative sort; rather than build its own memorial, the new state simply repurposed a structure built by the old state, Gothic architecture and Christian cross notwithstanding.

The government went a step further in 2010, in the run-up to the much-touted Commonwealth Games, illuminating the memorial in a bright red light. The lighting had an unintentionally ghoulish quality, bringing into stark contrast the banality of the government's tourist-minded plans versus the bloodshed and imperial excess the memorial originally signified. When asked about the appropriateness of celebrating British colonial sites in preparation for the Games, Delhi's Chief Minister Sheila Dixit gave a pragmatic response: "History is history and now we're talking of today: tourism, cultural ties and common links."[21]

Refuge and Rupture on the Southern Ridge, 1857–1947

Against Dixit's vacuous, business-minded optimism stands the more realistic assessment of William Faulkner: "The past is never dead; it's not even past." Surely India's past is all too alive in its present. To take the most obvious example: the wounds of Partition are, in many ways, still fresh. While the Northern Ridge, with its Mutiny Memorial, remains a testament to colonial oppression and rebel resistance, the Southern Ridge, and especially the Sufi shrine of Qutbuddin Bakhtiar Kaki in the present-day neighborhood of Mehrauli, offer a different history lesson: of a composite culture slowly slipping away, of antagonisms between Hindu and Muslim becoming ever sharper, and of state violence turning inwards to consume a shrine, a city and an entire subcontinent.

Qutb Sahib's shrine was deeply symbolic of the syncretic culture over which the late Mughals had presided. As the temporal power of the Mughals waned, they became increasingly devoted to religious activities. Zafar, the Mughal Emperor who became the unlikely figurehead of the Uprising, was especially fond of Qutb Sahib's shrine. His father had built a palace next to the shrine, and Zafar expanded the structure, adding an ornate red sandstone façade and many other elegant flourishes. He used it as a summer house, now rechristened "Zafar Mahal", a place to enjoy the relative coolness and breeziness of the Southern Ridge.

The shrine was visited by Muslims, Hindus, Sikhs and Christians, and it had been one of the focal points of a remarkable event, Phool Walon Ki Sair (the Flower-Sellers' Stroll), which had begun during the rule of Akbar Shah II, Zafar's father. The event has an elaborate backstory. Akbar Shah II had strained relations with Zafar, his eldest son, and planned to make his younger son, Mirza Jahangir, heir apparent. The British Resident in the Red Fort, nominally in the service of the Emperor but in fact the real source of authority, rejected this plan. Soon after, an incensed Mirza Jahangir, perched on a roof in the Red Fort, fired his gun at the Resident, missing him but killing his orderly. Mirza Jahangir was exiled to Allahabad, but his mother intervened on his behalf, pleading with the Resident and praying to Qutb Sahib, promising that she would pay her royal respects at the shrine if her son was freed.

Remarkably, the Resident gave in to the mother's demands, and allowed Mirza Jahangir to return to Delhi. The mother kept her word, and, along with the Emperor, took the 11-mile journey from Red Fort to Qutb Sahib's shrine, laying a garland of flowers on his tomb. In one local retelling of the tale, the Queen had decided to walk barefoot, and the flower-sellers, who were whole-heartedly devoted to the Mughals, walked with the royal couple, throwing flower petals on the road to soften the hardship of the long walk. Along the way, the flower-sellers, many of whom were Hindu, told the Queen that they too had been praying for the release of Mirza Jahangir, and that she should pay her respects at the nearby Yogmaya Temple as well. She happily did so.[22]

This retelling paints far too rosy a picture of the royalty and their unquestioning, loving subjects, but it does indicate the attitude of religious inclusiveness that pervaded the late Mughal era. The success and popularity of the Queen's dual pilgrimage led to the establishment of Phool Walon ki Sair as an annual event, expanded to include wrestling matches, fireworks, dances, swimming competitions and a full week of general revelry. The festival reached its peak during the reign of Zafar. A generous patron of the arts who was well-known for his interest in Hinduism as well as Sufi mysticism, he threw his energies into cultural and spiritual pursuits, with the Phool Walon ki Sair as the most public, celebratory expression of this.

This all came to a stop with the Uprising. Early in the morning on 17 September 1857, when it became clear that the British were going to take the walled city of Delhi, Zafar slipped out of Red Fort with a small party of attendants. His destination was Qutb Sahib's shrine; he stopped about mid-way at Nizamuddin's shrine to pay his respects and gather his energy. As he set out for the Southern Ridge, though, he was intercepted by a cousin, who told him that Gujjars were robbing all the people who passed that way. Though this explanation was likely true, it was a subterfuge; the cousin was actually in the employ of the British and had come to convince Zafar to surrender. He was successful. Zafar gave himself up to the British at Nizamuddin's shrine, and he was exiled to Burma. The grave intended for him outside Qutb Sahib's shrine lies vacant.

In the immediate aftermath of the British-led massacres in Delhi, Qutb Sahib's shrine, long considered a sanctuary for the embattled, assumed this role in a grim, literal sense. Refugees from the walled city streamed into this safe haven. But they had little food and were hardly prepared for the coming winter. The poet Ghalib, even after witnessing all the barbarities of the Uprising, was still shocked at the utter callousness of the British towards the families that had been driven out of the city and were wasting away at places like Qutb Sahib's shrine. He lamented, "Are the British officers not aware that many innocent and noble-minded women, both young and old, with small children are roaming the forests outside Delhi?"[23] Sickness soon broke out at Zafar Mahal, the summer palace adjoining Qutb Sahib's shrine, now jam-packed with refugees. In November, the British started letting Hindus back into the city, but Muslims were not allowed in unless they obtained a special order from the British government.

Tensions between Muslims and Hindus certainly predated 1857, but the defeat of the Uprising exacerbated these tensions significantly. The early days of the Uprising had seen an unprecedented level of cooperation and amity between Muslim and Hindu rebels. As the threads of the rebellion unraveled, the inter-religious bonds forged in early battles began to weaken. The British, as always, were eager to exploit these divisions, and after regaining control of the subcontinent, they took up their policy of divide and rule with renewed vigor, with special ire reserved for Muslims.

Remarkably, the Phool Walon ki Sair was revived in the years after the Uprising, and it continued through the late nineteenth and early twentieth centuries. But the British banned it in 1942, at the height of the "Quit India" movement. During this tense time, British officials feared any large gathering of Indians, especially one which threatened to bring together both Hindus and Muslims. The British sought to suppress the movement and succeeded largely by imprisoning the bulk of the Congress leadership. But by the end of World War II, the British realized that the independence movement could not be contained so easily. Two years after the end of the war, India gained its independence, but at a terrible cost: the trauma of Partition, which rent British India in two and led to untold suffering and bloodshed.

With Partition, Delhi was rocked by the worst violence it had experienced since 1857. Muslim-dominated areas, including the areas around Zafar Mahal and Qutb Sahib's shrine on the Ridge, were targeted by roving Hindu mobs. Once again, localized violence on the Ridge was a symptom of much larger political and military ruptures.

This was not the violence of the colonial empire against the rebelling colonized, as in 1857, nor of invading armies, as in the case of Timur or the Ghurids; these were two colonized communities tearing each other apart, just at the moment of their supposed freedom. Perhaps this is one reason why, in postcolonial India, anti-British sentiment rarely reaches the fever pitch of anti-Pakistan sentiment; the founding violence of the current Indian state was fratricidal, rather than directed at the colonial enemy.[24]

During Partition, the shrine of Qutb Sahib once again took on special significance. From its centuries-old role as a site of pilgrimage for Hindus and Muslims alike, it went on to become a victim of communal fury. Gangs of Hindu men from nearby villages ransacked the shrine, driving away all the devotees, damaging the stonework, and stealing the metal foil which covered some of the ornamentation (which they mistakenly thought was silver).

Gandhi had largely been sidelined during the negotiations between Nehru and Mountbatten. Though earlier, he had seemingly voiced his support for Partition (writing, in 1928, "I am more than ever convinced that the communal problem should be solved outside of legislation, and if in order to reach that state, there has to be civil war, so be it"[25]), when it came into effect, he was horrified by the religious violence it sparked. He spent the last months of his life traveling to affected areas and using his considerable mass appeal to try to contain the violence.

The last fast of his life was prompted by the ransacking of Qutb Sahib's shrine. This was the catalyst for a fast that eventually had several demands, all of them focused on ending communal violence in Delhi and beyond and ensuring that Muslims who had been chased out of the city could come back. From 13 January to 18 January 1948, Gandhi fasted, growing progressively weaker as the days went on. Finally, on the 18th, top political leaders, including members of Hindu nationalist groups like the RSS and the Hindu Mahasabha, signed and delivered a pact promising that Muslims would not be targeted in Delhi, and specifically that Qutb Sahib's shrine would be repaired and returned to its caretakers. Twelve days later, Gandhi was assassinated by Nathuram Godse, a Hindu nationalist who felt Gandhi was being too kind to Muslims.

Nehru resurrected the Phool Walon ki Sair in 1962, in a belated attempt to mend the communal fabric that had been torn apart during Partition. Later, Indira Gandhi tried to take the festival in another direction, inviting cultural troupes from each state to perform during the celebration. This was part of a larger effort to promote what was euphemistically called "national integration": in essence, a centralization imposed from above (as Indira Gandhi had tried, ultimately unsuccessfully, with the Emergency), with local autonomy replaced with caricatured cultural performances from different regions, thus creating a facade of diversity.

But the sheer persistence of the Phool Walon ki Sair shows both the reach and the limits of state power. On the one hand, the state has been instrumental to the founding of the festival and its many iterations. And the symbolism of Qutb Sahib's shrine goes back to the early days of Delhi as a political capital, as Qutbuddin Bakhtiar Kaki's spiritual power faced off against Iltutmish's political power.

On the other hand, the festival has consistently escaped the limits that the state has set for it. The festival is, after all, named for the flower sellers, not for any prince or prime minister, and its energy largely came from the impromptu shops and stalls that sprung up during the weeklong festival, selling sweets and kebabs and kites and jewelry and much else. Even today, more people attend the adjoining fair, with its ferris wheel and haunted house and myriad games, than the lackluster cultural performances. The people of Delhi, and particularly of Mehrauli, have been witness to, and sometimes, active participants in, state-sponsored violence, and they know its awful power. But the playfulness and frivolity of the festival suggests another kind of power, a resilience and lightness that escapes the long arm and stern gaze of the state.

PART II: STRUCTURES

Building State Power on the Ridge

As the Phool Walon ki Sair suggests, the state is not just concerned with military might. It also seeks to display its grandeur and magnanimity. Festivals are an ephemeral way to do this; more permanent is the construction of massive stone structures as monuments to state power. Take, for instance, Zafar Mahal: its sheer size and its employment of ornate stone carvings announce the prestige of its owner, but its location, adjoining a bustling Sufi shrine, far from the Mughal power center of Red Fort, implies a more humble, mystical bent. The palace suggests that rulers seek to make an impression not only through sheer power, but also through sophistication, luxury, spiritual devotion, pomp, and grace; that is, through majesty in every sense of the word.

As states have risen and fallen in Delhi, so too have a wide array of structures that were meant to convey the many shades of state power.

In some ways, the history of urban architecture in Delhi has been cyclical, just like the history of the city itself: empires have come and gone, buildings have sprung up and then fallen into ruin. But despite this ebb and flow, there has also been a linearity: the city has been getting bigger and bigger, from tiny Rajput fortresses to the immense urban sprawl of the present day. State efforts to structure the city have thus expanded and intensified, from austere walls to individual pillars and palaces to planned forests and overarching Master Plans to guide the growth of a megacity. And as structures have sprung up on all sides of an expanding city, the Ridge has remained surprisingly central, a key element in constructing state power.

Pillars of (Contested) Power on the Southern Ridge

In our survey of state architecture, we may start with the reign of Qutbuddin Aibak, who played such a crucial role in establishing Delhi, and more specifically the Southern Ridge, as a center of imperial power. Aibak initiated the construction of two iconic monuments on the lower slopes of the Southern Ridge, the Qutb Minar and the adjoining congregational mosque. Though Qutb Minar is Delhi's most famous monument, it is functionally merely an accessory of the nearby mosque and was used as a place for the muezzin to call the faithful for prayer. The mosque has several names, but it is most often referred to as *Quwwat-al-Islam*, or "Might of Islam". It is built using the remains of plundered temples, with Shaivite, Vaishnative and Jain columns interspersed throughout, and defaced idols staring blankly from the walls. Aibak began the construction of the mosque soon after Prithviraj Chauhan's defeat, siting it near a temple that was demolished in the fighting. The name of the mosque, "Might of Islam", was a natural choice, given that its main purpose was to strike at the religious heart of the heathen Hindu natives and to impress upon them the wrath of a new power boldly asserting itself.

Or so the story goes. This conventional interpretation of the Qutb Minar and its mosque is deeply ingrained. It is the story one gets in textbooks, in scholarly journals, in popular retellings, even on guided tours of the Qutb complex, now a UNESCO World Heritage site. But it is quite misleading. While it is true that Aibak and other Sultans destroyed temples, this is not necessarily evidence of an eternal Muslim/Hindu struggle. When Hindu rulers fought each other, they would also destroy their rivals' temples, so the mere act of temple destruction need not imply implacable religious enmity. Moreover, the standard narrative overstates the strength and unity of the new rulers.

As the historian Sunil Kumar has pointed out, no inscriptions in the Qutb complex refer to the mosque as "Might of Islam" and no Persian chronicles from this period contain that name. Kumar has taken great pains to dismantle the standard narrative of this site.[26] He notes that the name "Might of Islam" was first recorded only in 1847, some 750 years after the mosque was built, as part of a monograph on Delhi's topography written by Sayyid Ahmad Khan, an Indian judge employed by the East India Company. The name stuck, largely because it aligned so well with the British "divide and conquer" policy, with its stress on the irreducible difference between Muslims and Hindus in India.

But a closer look at the Qutb Minar and its adjoining mosque shows that they are not eternal symbols of Muslim fanaticism and Hindu defeat. Though, centuries later, they appear transfixed, as stark, unchanging symbols of a decisive victory, their actual history is full of change, uncertainty and transformation, reflecting the fluctuations of state power in Delhi. Just as there is no "pristine glory" in the ecology of the Ridge, there is no fixity to these Ridge structures and the states that built them.

The original minaret, build by Aibak, was quite short and squat, only the first storey of the present-day tower. There is no evidence that Aibak intended to build a taller tower, though it is easy to forget this and see the seeds of the contemporary minar in Aibak's creation. Built immediately after Ghori's victory over Prithviraj Chauhan and Aibak's seizing of the town from Ghori's chosen vassal, the minaret and the adjoining mosque are full of inscriptions that emphasize military might and victory over the infidels. The inscriptions sing Ghori's praises, with Aibak clearly intending the glory of his master to reflect on him.

Many histories have jumped to the conclusion that these inscriptions were aimed at the non-Muslim population that Aibak had subjugated. This certainly fits with the idea of a united Muslim ruling class imposing its will. But Sunil Kumar argues that Aibak had another audience in mind. The inscriptions are written in Arabic, in a script that would only be understood by Muslims with extensive education. Remarkably similar inscriptions can be found in the mosques constructed by Aibak's rivals, the other slaves-turned-sultans of North India. For example, his chief rival, Bahauddin Tughril, built a mosque in Bayana, which mirrors the Delhi mosque in its architecture and purpose, and which also contains inscriptions celebrating martial victories, divine favor, and the creation of a new congregational space for Muslims. Tughril's and Aibak's inscriptions should be read together, or rather against each other: each targeted, not the non-Muslim conquered, but the Muslim subjects of the new Sultans, who were competing against each other to prove that they had the strongest claims to sovereignty.

Further additions to the Delhi minaret and mosque reveal more twists in the tale of rivalry and contested state power. As we have seen, there was considerable turmoil after Aibak's death, and his eventual successor Iltutmish spent decades fighting both internal and external foes. After he finally consolidated his power, Iltutmish added three more stories to the minaret and expanded the congregational mosque significantly, doubling its original size. His mosque now dwarfed those built by erstwhile competitors like Tughril, and the minaret now towered over the landscape. The inscriptions he added to the mosque are also revealing; although some speak of military victory, the majority of inscriptions are focused on building a properly devout community, exhorting Muslims to congregate and pray at the mosque—in keeping with his promotion of Delhi as a refuge from the predations of the Mongols.

These monuments of piety, pride and power took on new significance with the rule of Alauddin Khilji, whose father had toppled the already teetering Slave Dynasty and taken control of the Delhi Sultanate. Khilji's additions to the mosque and minaret show his ambitions and his hubris. First, he doubled the size of Iltutmish's already-expanded mosque. Though little of this extension remains today, a glimpse of it can be seen in the enormous, elegant entrance hall on the south wall of the mosque. The hall is made of red sandstone, with elaborate marble trim and inscriptions from Koranic verses. Now standing alone, the hall was once a fittingly grand entrance to an enlarged and newly adorned mosque. In contrast, a rougher piece of architecture sits outside the mosque, the beginnings of a new minaret that Khilji intended to be double the height of the existing minaret built by Aibak and Iltutmish. But Khilji could not even finish the first storey before his death; all that remains is the jagged rubble masonry of the core of the minaret, which Khilji planned to cover with marble and sandstone.

The next major intervention in the architecture of this complex came with Firoz Shah Tughlaq. Given the attention he had lavished on Ashokan pillars, it is perhaps not surprising that he had a keen interest in Aibak and Iltutmish's minaret as well. In 1368, the minaret was struck by lightning, which sent the top storey flying off. Tughlaq did not let this opportunity go to waste; he replaced the destroyed storey with two storeys of his own, adding large amounts of white marble to differentiate his addition from that of his predecessors.

By the time that the British and their allies began writing about the Sultanate-era mosque and minaret on the Southern Ridge, centuries had passed since this last intervention, and the moral geography of the area had shifted considerably. This brings us back to Sayyid Ahmad Khan, the East India Company judge who first recorded the use of the name "Quwwat al-Islam" (Might of Islam) for Aibak's mosque. Given the history of the mosque, especially the trauma of Mongol invasions, it becomes clear that "Quwwat al-Islam" is in fact a corruption of "Qubbat al-Islam" (sanctuary and axis of Islam), as the latter is a well-documented name for Delhi. But for Khan, and especially for the British historians who followed in his wake, this history was pushed aside, and replaced with a picture of unending Hindu-Muslim conflict.

In a similar way, the name of the minaret, Qutb Minar, was taken to refer to Qutbuddin Aibak, the military strongman who had imposed his will on the native population. But the name was of relatively recent vintage, and in fact referred to the Sufi saint, Qutbuddin Bakhtiar Kaki. In local tradition, Kaki had risen to the top of the Sufi hierarchy, and was thus considered to be the axis of the world, centering and stabilizing existence through his mystical presence. The minaret became known as Qutbuddin's staff; it was a representation of the saint's power to connect heaven and earth.

But these local meanings were swept away by the heavy hand of British historiography. The British emphasized the military, and orthodox Muslim, origins of the mosque and the minaret. They also sought to make their own imperial mark on Qutb Minar, repairing it in the aftermath of an earthquake in 1803, and, at the behest of one Major R. Smith of the Royal Engineers, replacing the cupola at the top with a structure modeled on a Bengali pavilion. But the jarring dissonance of this new architectural form was too much for even the British to take, and they removed it in 1848, dumping it unceremoniously on the grounds near Qutb Minar, where it remains till today.

Location, Location, Location (the Art of Siting a New Capital)

By the twentieth century, British ambitions for Delhi had grown well beyond tinkering with Southern Ridge monuments. In 1911, the British announced their intention to move the Raj capital from Calcutta to Delhi. Building the new capital took 20 years. Once it was finally inaugurated, the British could only enjoy their new capital for 16 years, before Indian Independence swept them away, yet another lesson in the impermanence of state power.

Although the British were ushered out of the capital, and the country, in 1947, their influence still permeates the city that they built, and well beyond. British officials consciously sought to link themselves to empires of old; but at the same time, they were clear that they were building a new city on new principles, including new legal frameworks, new aesthetic sensibilities, new urban planning tools and new ecological visions. And, despite the upheavals of Independence and Partition, these are still, largely, the principles that govern Delhi today, and that determine the fate of the Ridge.

The transfer of the British capital from Calcutta to Delhi was then, a momentous turning point for the city of Delhi and its Ridge, one which depended on its past prestige but charted out a radically new course for its future. The initial 1911 announcement was largely seen as a punishment for Calcutta and the larger Bengal region, which had been at the forefront of resistance to British rule. But the pull of Delhi should not be underestimated.

In a letter written about a month before the transfer of the capital was publicly announced, Lord Crewe, the Secretary of State for India, noted that Delhi had "an Imperial tradition comparable with that of Constantinople, or with that of Rome itself." In addition to the symbolic legacy that the Sultans worked so hard to establish, and which the Mughals readily embellished, there were older, more mythical resonances. Crewe invoked the *Mahabharata* and the city of Indraprastha, noting that "the near neighborhood of the existing city formed the theatre for some notable scenes in the old-time drama of Hindu history, celebrated in the vast treasure-house of national epic verse."[27]

For Crewe, and for many other British officials, the Uprising of 1857 could only be explained as a burst of irrational passion, with the barbarous natives showing undue attachment to old lineages and centuries-old symbols of power. Having failed to appreciate these sentiments before, the British were now willing to exploit them however they could. Crewe continues:

> To the races of India, for whom the legends and records of the past are charged with so intense a meaning, this resumption by the Paramount Power of the seat of venerable Empire should at once enforce the continuity and promise the permanency of British sovereign rule over the length and breadth of the country.[28]

Of course, the British had also built up their own attachment to the "legends and records" of Delhi, especially in the wake of the Uprising, as the Mutiny Memorial on the Northern Ridge suggests. Indeed, the historical significance of Delhi, and especially of the Northern Ridge, made the choice of the new capital appealing, not just to British officials in India, but also to a wider British public that had been raised on tales of imperial derring-do and military triumph.

Initial speculation in the press was that the new British capital would be built in close proximity to the Northern Ridge. And indeed, this site was thoroughly scrutinized by the three-member Delhi Town-planning Committee, which was formed to evaluate potential locations for the new city and to guide its planning. But in the end, the Committee decided against siting the new city around the Northern Ridge. In part, the very sanctity of this site worked against it; the Committee reported that,

> sentiment will not permit of new buildings being erected on the better half of it. The portion from Flagstaff Tower to Hindu Rao's house and the Mutiny Memorial must remain sacrosanct; the Ridge can therefore never be more than a rough park garnished with plain but hallowed buildings.[29]

Sentimental concerns were not the only ones, though; the potential northern site for New Delhi, which included both the Northern Ridge and the plains running down to the Yamuna, was rejected for a number of reasons: it was too small, not allowing for future expansion of the city; the plains were often waterlogged and malarial; and the area was already occupied by many British civilians, whom it would be difficult (and expensive) to displace, especially because, as the Committee noted, a majority of those civilians "represented the business houses of Delhi".[30]

In the end, the town planners abandoned the north, but they did not abandon the Ridge; the site they chose was centered on Raisina Hill, in a zone the British referred to as the "Southern Ridge". This was the first time the English term "Ridge" had been applied beyond the Ridge of 1857 fame, and it marks the slow transition of the term's meaning, from its original limited sense (the ridge north of Shahjahanabad), to its expanded, generic sense (all of Delhi's hills). The names of particular sets of hills have changed as the city has expanded. What the British called the Southern Ridge, around Raisina Hill, is now called the Central Ridge, which is a significant appellation both geographically (it's now in the middle of the city) and symbolically (it's at the heart of the present-day iteration of Delhi).

In later writings, the Raisina site was freighted with the weight of inevitability; its choice as the locus for the new city was portrayed as pre-ordained. The Governor-General of India at the time, Lord

Hardinge, later wrote of a horse ride he took up to Raisina Hill with William Hailey, the Chief Commissioner of Delhi:

We galloped over the plain to a hill some distance away. From the top of the hill there was a magnificent view, embracing Old Delhi and all the principle monuments situated outside the town, the Yamuna winding its way like a silver streak in the foreground at a little distance. I said at once to Hailey: "This is the site for Government House", and he readily agreed.[31]

One of the chief architects of the new city, Herbert Baker, describes it more lyrically, almost mythically. He was standing on Raisina Hill with two friends, looking at

the deserted cities of dreary and disconsolate tombs and wondering how the new city would rise. The sky was overcast and it rained intermittently. Suddenly, the clouds lifted and the sun broke through. A brilliant rainbow formed a perfect arch on what was destined to be a great vista, where Lutyens' memorial arch [India Gate] now stands. We acclaimed it was a good omen.[32]

Omens aside, the British were clearly taken with the view from Raisina Hill; once again, the sheer height of the Ridge came to play a key role in state decision-making. Whereas earlier empires (and even the British in 1857) had been drawn to the heights of the Ridge for military considerations, the British town planners were guided by more aesthetic reasons. But aesthetics were not divorced from power; the impressive views from Raisina Hill were employed to emphasize both the continuity of British and pre-British empires and the supremacy of the new rulers.

This was stated quite explicitly by George Swinton, head of the Delhi Town-planning Committee:

The British Raj has come up at last to range itself alongside of the monuments of past rulers, and it must quietly dominate them all.... It is because I find it difficult to see that expression of dominion in what I fear may develop into little more than a superlatively well arranged cantonment that I have personally looked to the rock and the "command" of the Ridge.[33]

From the Ridge, the British rulers could literally look down on the earlier cities of Delhi. Raisina Hill formed the focal point of a semi-circle of historic monuments, with the Mutiny Memorial to the north, Shahjahanabad's Jama Masjid to the northeast, the tomb of the Mughal Emperor Humayun to the east, and the tomb of the Mughal statesman Safdarjung to the south.[34] In early plans for the new city, the link between Raisina Hill and Jama Masjid (between what was increasingly being called "New Delhi" and "Old Delhi") was emphasized; early sketches included a grand boulevard between the two sites. The new city would be happy to soak in the prestige of the old empire, but with a firm emphasis on its dominance.

In the end, though, the plan for the grand connecting boulevard was dropped, largely due to cost concerns. This suggests that, for all their emphasis on grandeur, town planners had to consider other factors as well. British officials were constantly torn between the need to make "New" Delhi suitably impressive and the importance of keeping costs at a reasonable level, especially as critics in India and back in Britain began to decry the extravagance of Raj expenditures.[35]

From an economic point of view, Raisina Hill was a good choice. Unlike in north Delhi, there were no wealthy business owners who would have to be compensated. And, largely because of its rockiness, the land itself was cheap. An initial report on potential costs of land acquisition noted that "waste" land generally had a market value of around ₹15 or ₹20 per acre, or up to ₹40 per acre closer to Shahjahanabad, where "grazing is valuable". In contrast, fertile farming land, especially cultivable land close to the city, was valued at up to ₹140 per acre. (Of course, these seemingly objective valuations were only possible because of major transformation of the land which the British themselves had set into motion, including the treatment of land as a mere commodity and the marked preference for agriculture at the expense of pastoralism, as detailed in the previous chapter.)

Raisina Hill had another important benefit, in addition to the economic and the aesthetic: its slope kept it relatively free from waterlogging and flooding, and thus reduced the risk of diseases like cholera and malaria. The Delhi Town-planning Committee attached great importance to the fact that, in this area, *storm water and drainage of all kinds can have a rapid fall and be given an outfall above flood level.... The investigations of the sanitary officers prove that the villages in this area have the healthiest past history of any of the areas under the consideration of the Committee.*[36]

The Committee had one more factor in mind when considering the Raisina Hill site, which should come as no surprise given the history of state intervention in Delhi and on the Ridge: the role of the military in the new city. In keeping with the projected scope and scale of the new Raj capital, government officials planned for a grand new army cantonment in Delhi. The cantonment would ideally be close, but not too close, to the key government buildings of the new city, so that the civil and military authorities could be in close communication but would also have their autonomy. As the civilian planners were closing in on a site on the eastern slopes of the Southern (now Central) Ridge, they kept up a dialogue with military authorities, who were increasingly interested in the plot of land on the western slopes of the same portion of the Ridge. Whereas the eastern site would gaze out at the historic cities of Delhi, the western site would spread out onto a more open, empty plain, ideal for military training exercises and further expansion of the cantonment.

In short, the Raisina site, due in large part to the characteristics of the Ridge, met all of the imperial requirements for a new city, including aesthetic, economic, sanitary and military concerns. The confluence of these factors, quickly identified after a cursory investigation of possible sites, quickly ended the debate about the siting of the new city. Raisina Hill would be the focal point of Delhi.

In many ways, it still is. The governmental heart of Delhi, and indeed of India, is still located on a small rise in the Ridge; "Raisina Hill" has become a metonym for state power, akin to "The White House" in the United States or "10 Downing Street" in Britain. The zones surrounding the (now Central) Ridge, on both its eastern and western slopes, still have special status. While most municipal concerns in the city are addressed by the Municipal Corporation of Delhi, two privileged zones have their own caretakers: Raisina Hill and its surrounds are looked after by the New Delhi Municipal Council, while the military has its Cantonment Board.

Not that these seats of civilian and military power have gone totally unchallenged, both by civil society and by other parts of the government. The tension between local government bodies and imperial (and then national) central governments has simmered in Delhi since 1911, as the quarry quarrel of Chapter 1 amply demonstrates. Central to that quarrel was Major Henry C. Beadon, the Deputy Commissioner of Delhi District. But Beadon's role in the creation of New Delhi was not limited to impeding quarrying efforts and insisting on a proper chain of command. He was, in fact, central to the process of acquiring land for the new capital. Without his help, New Delhi would have no ground to stand on.

The Travails of Land Acquisitions, on Raisina Hill and Beyond

From government correspondence, it is clear that other officials found Beadon to be insufferable but at the same time indispensable. Choosing a site for New Delhi turned out to be a relatively simple task, but actually building a city around Raisina Hill was a gargantuan undertaking. And before any construction began, land needed to be acquired. For this, the government needed someone with an intimate knowledge of the area and its land use customs and regulations. So, they turned to Beadon, author of the 1910 settlement report for Delhi.

Government officials knew how high the stakes were, and how costly a wrong move could prove to be. Even before the transfer of the capital, land speculation was already widespread in the Delhi area. Officials recognized that, once Delhi became the new imperial capital, the land market would go into a frenzy and land prices would skyrocket.

In an attempt to rein in the effects of land speculation, officials made careful use of the Land Acquisition Act of 1894. In many ways, this law built on the logic of the Indian Forest Act of 1878, which had been introduced to give the British control over the no-longer-useless "wastes" of India. The Land Acquisition Act was much more sweeping, establishing procedures for the government to forcibly acquire any land it deemed necessary for "public purposes". The language of the act was deliberately vague, giving the government considerable leeway to determine what exactly constituted a "public purpose" and giving Indians very little recourse if the government decided it wanted their land.

Within two weeks of announcing Delhi as the imperial capital, 115,000 acres (roughly 465 square kilometers) for the new city had been notified under the Land Acquisition Act. This initial notification was largely a preventive measure, as the site for the new city had not yet been selected. But according to the Land Acquisition Act, compensation rates were tied to market rates for land on the day of the initial notification (even if the actual payment for the land happened much later). British officials realized that they were about to make land in Delhi very, very valuable, and they wanted to artificially freeze the prices so they could buy up land at their leisure. Despite their ideological embrace of free market forces, they were happy to disrupt these forces when they became inconvenient.

The Delhi Town-planning Committee submitted its report recommending Raisina several months after the price freeze, in June 1912, and by August 1912, the ever-efficient Beadon had produced a report detailing the estimated costs of acquiring the plot of land the Committee had recommended. Though significantly less than the originally notified area, it was still enormous: roughly 43,000 acres, along with an additional 9,000 acres tentatively set aside as an "Expansion Tract". Beadon's report records, in painstaking detail, the different types of land to be acquired, and arrives at the final figure of ₹373,060 as the tentative cost of acquisition.

But despite the seeming precision of Beadon's report, all was not well in officialdom. Before submitting the report, Beadon, as usual, was complaining about the shortcomings of the people and systems impeding his progress.[37] The city also proved surprisingly difficult to map. As the earlier Settlement Reports of Beadon and others suggest, the colonial state got its power, not just from military might, but from its abilities to draw boundaries, create categories, and map its territories with increasing precision; this enabled a range of new taxes and regulations, which changed the use of land and altered the growth of villages and towns in dramatic ways.

But this was easier said than done. Officials found it particularly difficult to find suitable maps for the areas around the Yamuna River, whose water levels fluctuated with the seasons, and whose path had shifted over the years. Demarcating this area for the capital was a comedy of errors, with lost maps, indignant engineers, misplaced boundary markers and suspicion of sabotage. It took the beleaguered officials several years to finally draw the official boundaries for this part of the city.[38]

Once the land was mapped and demarcated, even if inadequately, the acquisition could be finalized. But this too was no easy process. In Beadon's opinion, the problem was that the Land Acquisition Act was too lenient. In his (typically patronizing) words, the Land Acquisition Act

> has the blemish that it does not recognize the custom of the people of this country to be dissatisfied with any decision which is not the pronouncement of the highest tribunal.... It will be necessary to employ a full time advocate to cope with the litigation.[39]

It takes a particularly blinkered colonial viewpoint to see the Land Acquisition Act as a lenient law. In Delhi, the Act has been a tool for forced urbanization, liquidating agricultural lands, grazing grounds and village commons and replacing them with government buildings and urban infrastructure. Officials knew that, with the growth of New Delhi, life in the region would change irrevocably; but the British had little interest in giving villagers a stake in the new city.

Freezing "official" land prices in late 1911 using the Land Acquisition Act was the first step in villagers' marginalization, assuring that they would not benefit from the bonanza of rising land prices in Delhi. But this was just a prelude to their wholesale dispossession. Entire villages were uprooted to make way for the new city. Raisina Hill, for instance, gets its name from Raisina Village, whose inhabitants were forced to abandon their lands, including their old village commons on the Ridge.

Sometimes, in lieu of (inadequate) compensation money, villagers were given land elsewhere, so they could, in theory, continue their livelihoods and put their agricultural skills to good use. But the government's relocation efforts were half-hearted at best. Originally, officials promised farmers plots of fertile land near their old villages.

It quickly became evident that such land was near-impossible to find, and more far-flung spots were identified, including "wastelands" in Karnal and Rohtak. In another scheme, farmers were given plots of government-owned land in Punjab, but only as tenants, with eventual ownership of the plots dangled as a distant possibility.

Even after employing these dubious tactics, the acquisition of land for New Delhi was still an expensive proposition, which exceeded Beadon's original estimates. As the city planning moved from the initial cost estimates to the actual payment of compensation to land-owners, Beadon recedes in the archival records, and is replaced with an Indian civil servant named S. S. Khazan Singh. Given the title of Special Land Acquisition Officer, Singh was tasked with the delicate job of negotiating with landowners who were about to be displaced from their ancestral villages.

Just like the contemporaneous Quarry Quarrel, Singh's acquisition efforts reveal the ambiguity of Indian officials in the British Raj. On the one hand, the fact that Singh was tasked with such a crucial responsibility suggests the Raj's increasing recognition of the skill and competence of Indian officials. On the other hand, Singh was clearly carrying out an imperial agenda, one which put him at odds with many of his compatriots, who were being uprooted by that very agenda.

Singh's Indianness likely helped him in his negotiations. He could relate to landowners in a way that the imperious, patronizing Beadon could hardly hope to. In his final acquisition report, Singh notes that "the task of acquiring this property was very heavy and intricate", especially because "acquisition operations [began] so late after the issue of the Gazette Notification of December 1911". Despite this, Singh was convinced that "the amount of compensation paid on the whole... shows the sum awarded to be a very moderate market value for the property acquired".

Further, the legal troubles that Beadon feared largely did not materialize, at least in Singh's account, which notes:

The reason that in case of such important acquisitions there should have been so few [legal] applications is that, in each individual case, I have been trying to convince the persons interes̶t̶e̶d̶ ̶o̶f̶ ̶t̶h̶e̶ adequacy of the award made. The procedure is in no doubt very laborious, tedious and troublesome requiring a considerable amount of tact and patience, but it is exceedingly beneficial to both the people and the Government, as it leaves no ground for the persons interested to grumble and saves the expenses and worry of litigation which otherwise is sure to be seriously large.[40]

The government, however, followed up on Beadon's suggestion and hired a lawyer from Lahore to work full-time dealing with legal challenges to the government's acquisition and compensation decisions, suggesting that the process was not quite as smooth as Singh implies.

An Acquisition Postscript

There is more recent evidence, too, that Singh's claims may have been overblown. A century after the massive acquisition process began, its after-effects are still rippling through the Indian court system. At least 17 cases have been brought against the government in recent years, from families who claim that their relatives were never properly compensated.

Admittedly, there was a long gap between the original acquisition and the filing of these court cases. In the intervening years, anger at the British was channeled, not through official legal channels, but through village lore and long-nursed grievances. Most of the current legal challenges have originated from a small village named Malcha Patti in Sonepat district of Haryana, roughly 50 kilometers from New Delhi. The place gets its name from a much older Ridge-side village called Malcha that bordered Raisina and was completely obliterated to make way for the grand government buildings and boulevards of New Delhi. The current inhabitants of the Haryana Malcha are descendants of the Delhi Malcha, villagers who were driven out by British machinations.[41]

Some in the Haryana village remember hearing stories from their grandparents about British troops storming the villages of Raisina and Malcha, and killing 33 villagers protesting their impending displacement. Despite the specificity of this memory, it is difficult to verify. The British, though hardly averse to celebrating their massacres, left no record of a military action to clear these villages.

British officials did, though, keep meticulous records of land owner-ship and government payments. After determining what they felt were fair compensation rates for land acquisition, the government depos-ited the compensation money with the Court of the Divisional Judge in Delhi. The problem was that many of the villagers did not know how to claim the money. Some had already fled Delhi, embarking on new lives in areas like (new) Malcha, Haryana. Many were illiterate and had no way of understanding the compensation procedures. Their descen-dants now estimate that only ten of the 300 families in (old) Malcha and Raisina claimed their compensation money.

So where is that money now? This is the question that has hounded a rose farmer named Sajjan Singh for the past 20 years. Singh is an inhabitant of the Haryana Malcha, and his grandfather, now in his 80s, is one of the elders who has passed down stories of British cannons arriving in the Delhi Malcha. Singh is the reason that these stories have moved from the realm of lore to that of legal challenges. Back in 2006, Singh, having learned from a chance encounter with a Revenue Officer in Delhi that his forefathers should have been compensated for their land, hired a lawyer to track down the missing money.

If the British court had indeed deposited the money, it would have been in the official British bank: Bank of Bengal, which later merged with the Bank of Bombay and Bank of Madras to become the Imperial Bank of India. After Independence, it became the government-owned State Bank of India (SBI), which proudly proclaims itself "The Banker to Every Indian". In theory, SBI should now hold the money set aside for the Raisina and Malcha villagers.

But SBI officials, when questioned about this matter, responded in an affidavit, "The relevant records of the year 1913 of the Bank of Bengal are not traceable." SBI is not the only government issue to use this excuse. Sajjan Singh's lawyer laments, "Everyone is eventually hiding behind this one excuse—documents have been lost."

Faced with this implacable logic, Singh has been running around in circles trying to get the government to at least recognize that he and his family deserve compensation. Recently, he has focused on one particular plot of land that his family once owned, a 20-acre strip that has become embroiled in much larger controversy. The British government had leased the land to a Swedish entrepreneur named Edward Keventer for the purpose of establishing a dairy farm. Keventer's company was purchased by the industrialist R. K. Dalmia after Independence, and the Dalmia family kept renewing the lease for the plot, which was still, in theory, only to be used for dairy farming. Delhi Land and Finance (DLF), the largest real estate developer in India (and a major player in the next chapter of this book), bought the Dalmia's dairy company in the early 2000s, and they have since embarked on an ambitious plan to build high-rise luxury apartments on the plot of land.

DLF is known for its close ties to the Congress party, and several BJP notables raised questions about the propriety of DLF's high-rise plans in erstwhile Malcha. How, for instance, did they get the central government's permission to build eight-storey buildings in such a sensitive security zone? And why did they pay such a low fee to convert the official land use of the area from commercial (dairy) to residential (luxury apartments)?

As the main opposition party from 2004 to 2014, the BJP cried itself hoarse about Congress corruption, and Congress collusion with shady DLF plans was only one more instance of this. Through the network of contacts he had made during his decades-long legal crusade, Sajjan Singh began meeting with BJP officials who had become interested in his case as a way of targeting the Congress. Sajjan Singh says that he even met with Rajnath Singh, the upper-echelon BJP leader, who promised his full support.

But these promises disintegrated when the BJP came to power in 2014. Now that they had engineered the Congress' humiliating 2014 election defeat, they had little interest in going after the big corporations that had benefited from Congress patronage, lest they be held under similar scrutiny for their various corporate collusions. One BJP leader told Singh that the party "did not want to get caught in political thickets".[42]

From potential savior, the BJP went to being Singh's chief enemy, especially when it introduced a controversial ordinance to a key land acquisition law soon after coming to power. The law in question was the "Right to Fair Compensation and Transparency in Land Acquisition, Rehabilitation and Resettlement Act, 2013", passed by Congress as a replacement for the much-criticized Land Acquisition Act of 1894. However, the new act, while providing some additional benefits and safeguards for those threatened with displacement, keeps the same general framework of the British law, giving the government similarly sweeping powers to acquire land; it is telling that most news reports refer to the new act simply as the Land Acquisition Act of 2013.

The ordinance introduced by the BJP further weakened provisions meant to protect farmers and others dependent on the land. The ordinance inspired massive protests, which enjoyed support from a wide range of farmers' organizations, social movements and political organizations, and activists like Sajjan Singh. In response, the BJP quietly let the ordinance lapse when it came up for renewal.

But this minor victory hardly erases the damage done by many years of land acquisitions, first carried out by the British government, then carried forward by the Indian state. By the government's own estimates, almost 75 percent of the roughly 40 million people who have been displaced by government land acquisitions have not received compensation. And though the government will hardly admit it, very few of the remaining 25 percent have received adequate compensation; despite the government assertion that it buys land at above the market rate, farmers know that they are better off selling their land to private developers, even though such developers are notoriously unscrupulous.

Sajjan Singh is quite aware of these facts, as well as the staggering amounts of money that are involved in land politics. In the 1910s, government officials would have put the value of his forefathers' 20-acre plot at about ₹700; this is the compensation amount that has mysteriously disappeared. Even at the time, landowners complained bitterly about unfair compensation rates, but the skyrocketing land prices in Delhi, which rose far faster than inflation rates, make these compensation figures look truly ludicrous. The land which DLF now hopes to develop has an estimated market worth of ₹90,000 crores; Singh, in one of his many court appearances, said that he would thus consider ₹15,000 crores billion a fair compensation to his family now.

The court balked and said the government could hardly afford such an enormous amount. Singh responded that he would happily accept the land back instead; he would also accept a 100-acre plot of land in another, less lucrative part of the country. The court referred the case to the Lieutenant Governor of Delhi, the Land and Building Department, and the Ministry of Urban Development; Singh's Kafkaesque bureaucratic nightmare continues. Meanwhile, the former villages of the Delhi Ridge still make do on the plains of Haryana, as they have for the past hundred years.

Back to British New Delhi and its Ridge Transformations

The Land Acquisition Office set up by the British Raj was closed on 5 June 1916; as far as the government was concerned, the process of acquiring land was over. After all, they had a city to build, which meant they had a landscape to transform. As one account notes,

25 villages and their agricultural land had been acquired. The entire landscape of the land taken over was reworked, village settlements were levelled, many stretches of low-lying land were filled up, several hillocks were levelled and agricultural fields were replaced with macadamised roads and buildings.[43]

This included blasting 20 feet of rock from the top of Raisina Hill to make a level platform for the new government buildings.

As this description suggests, British planners had no concept of the Ridge as a "pristine" landscape to be protected. They were happy to chop down hills when it suited their purposes. And, as noted in Chapter 1, the British were busy quarrying other parts of the Ridge, with a dedicated rail line bringing quartzite blocks and Badarpur sand from the southeastern part of the Ridge to the central Ridge where New Delhi was being erected.

Despite their general lack of concern for ecological issues, the British laid the foundations for the modern Ridge forest when they built New Delhi. As detailed in the previous chapter, there had been sporadic British attempts to afforest the Northern Ridge, near the Mutiny Memorial, during the 1880s and 1890s. But these efforts, with little real backing from the government, sputtered out quickly. The afforestation of the Ridge only picked up real momentum with the founding of New Delhi.

The planning for this had started before the villagers of the Ridge had even been displaced. An early objection to the Raisina Hill site for New Delhi was that the surrounding Ridge was too barren, worn down as it was by years of grazing, urban pressures and short-sighted government land use policies. Although placing government buildings on a hill would certainly project power, the grace and majesty of the new capital would be undermined by the rough, dusty grimness of the Ridge's bare rocks. One government report notes,

From an aesthetic view, the ridge is not a pleasing sight. The crest of the ridge forms an unbroken hard line on the horizon, and its bare slopes littered with debris of rock, and its monotonous brown color due to the dry grass in winter render it unattractive.[44]

To counter this objection, Governor-General Hardinge sought the advice of the Conservator of Forests in the United Provinces, P.H. Clutterbuck, who assured him that the Ridge surrounding Raisina Hill could be reforested.

The first concrete step in the reforestation project was to stop grazing of the land. In the long-term, this would not be an issue, as all the villages in the area would be uprooted. But in the short-term, British officials went with their favored strategy of drawing boundaries, fencing off erstwhile grazing land and hiring guards to keep away village pastoralists. As opposed to the earlier, scattered attempts to afforest the Ridge, this one had the full weight of the imperial government behind it. To give the effort legal weight, the government notified the Raisina part of the Ridge as a Reserved Forest on 24 November 1913.[45] This was the first time, though hardly the last, that the Indian Forest Act would be applied in Delhi.

The language used by government officials at the time makes it quite clear that a landscape needn't actually be tree-filled to count as a "Reserved Forest"; rather, the label could be aspirational. Wherever the British (and later the Indian) government wanted a forest to be, it could proclaim a Reserved Forest. In place of the barren Ridge, where, at most, there were grasses and shrubs a few inches high, there would soon arise "a green mantle of vegetation".[46]

With the legal framework in place, and with the to-be forest fenced off, British planners saw quick results. Within a few years, many trees had sprung up. One forester noted that there were, in some places, "thickets it would be difficult to get through".[47] These were generally trees that thrived in the thin, rocky soil of the Northern Aravallis; they can still be found in places like Mangarbani.

But the government was not content with this regeneration, even though it had taken very little time and money on their part. From the beginning, the goal of the afforestation was aesthetic, and a rather specific aesthetic at that: the look of imperial grandeur. Not satisfied with "a scrub type of forest with trees of low height growth", city planners instead emphasized the need for "a better type of vegetation" (elsewhere: "a pleasing type of vegetation"), which would necessitate artificial watering.[48] Clearly, "better" and "pleasing" are highly subjective categories, and in this case, they were filtered through the lens of a thoroughly British landscape gardening aesthetic; the trees should be tall, straight and evergreen, and should produce sightly flowers.

But beyond the individual trees, the emphasis was the role that the Ridge would play in the larger landscape of New Delhi. As noted earlier, Raisina Hill was chosen as the center of New Delhi in part because it formed a promontory from which the older Delhis could be viewed and, quite literally, looked down upon. This function was not just confined to Raisina Hill, it applied to the Ridge as a whole. While planners were mainly concerned with the so-called "New Delhi Ridge", which surrounded the new capital, they did not forget the "Old Delhi Ridge", especially because of its Mutiny resonances. The latter received the "Reserved Forest" designation under the Indian Forest Act in 1915, two years after its more central counterpart.[49]

Connecting these two parts of the Ridge was a key part of constructing New Delhi and bridging it to Delhi's past, in both literal and symbolic ways. As the planning committee's final report notes:

The northern ridge must be considered also, for as soon as the linking road has been carried out, the drive along the crest from the Cantonments and Malcha to Hindu Rao's House and the Flagstaff Tower will become popular. Arrangements are now being made to protect its slopes, and, when to an unsurpassed sentimental and historic interest are added fine trees and shrubs and flowers, few places should have a stronger attraction....The views from these drives will be magnificent. The panorama of the present city, the new city and the monuments and cities of the past stretching below to the river as seen from the rough eminence past a foreground of rocks and trees should be one difficult to match for charm.[50]

The views were meant for wealthy British officials and civilians, who could afford the vehicles to drive along these new Ridge roads and who would appreciate the views of the decaying ruins along with reminders of the 1857 glory and the supposed role of the British as the redeemers of Delhi's long imperial history. And the backdrops of these views would be the Ridge, which thus could not be a mere "natural scrub forest", but instead had to engineered to be as majestic and stately as possible.

The "Scheme for the afforestation of the Delhi Ridge" drawn up in February 1913 by B. O. Coventry, Delhi's Deputy Conservator of Forests, gives detailed notes about how such a forest could be brought into being. It would not be enough simply to plant the desired trees during the rainy season; to achieve the desired aesthetic effect, the hills of the Ridge would need to be terraced, which would both create a uniform view and allow for soil regeneration. Coventry also contemplated more drastic measures, such as using dynamite to blast holes in the rock, which could then be filled with soil and planted with tall, water-hungry trees like the Australian eucalyptus.

Coventry realized that it would be difficult to create a lush, evergreen forest given the Ridge's unforgiving conditions; he notes that

owing to the peculiar nature of the ridge, with its shallow rocky soil, much of the work... will be more or less experimental, and experience alone will show exactly which species will succeed and which will not. There are many species, however, which are certain to succeed.[51]

Even this assessment was overly optimistic. Many of the supposedly "successful" species, for instance, were indeed evergreen in wetter conditions, but dropped their leaves during the dry season in the arid soil of the Ridge.

Reviewing the afforestation work six years after it had started, a Forest Department official named R. N. Parker recognized these flaws. After noting that "a more unpromising site for afforestation would be difficult to find",[52] Parker asserts that "many of the plants being planted on a big scale... are unsuited to the locality." He does though note that he found *Prosopis juliflora,* our old friend *vilayati kikar,* to be "very conspicuous".[53] Soon after Parker's initial visits to the Ridge, *vilayati kikar* found an enthusiastic proponent in William Mustoe, who was appointed Forest Officer for the Ridge in 1919.[54] Mustoe personally supervised trips around the Ridge to plant the persistent mesquite, ensuring its rapid spread around the Ridge.

Due to the unexpected (but ecologically unsurprising) proliferation of *vilayati kikar,* British afforestation efforts were successful in reaching their main stated goal: provide a suitably green backdrop to the new seat of imperial power. Since, from the beginning, the British plantation drive was driven by aesthetic and political motives, rather than ecological ones, officials were hardly worried about the potential impact of *vilayati kikar* on the diversity and long-term sustainability of the landscape.

In other aspects of this project, too, the British attitude was one of studied unconcern. Having given the initial notification for the two portions of the Ridge in 1913 and 1915, respectively, and having fenced off the areas and begun plantation efforts, the government began dragging its feet on the legal procedure mandated by the Indian Forest Act. This involved what the Act dubs a "Forest Settlement Officer" who was to survey the area and determine whether any compensation needed to be paid. The parallel with earlier "Land Settlement" programs, as well as ongoing land acquisition efforts, is striking: the goal was for the government to lay claim to land as quickly and cheaply as possible, with one lone officer left behind to settle any claims from locals whose lives were disrupted by the government's moves.

The initial surveying process must, according to the Indian Forest Act, be completed within three months. The officer then must review all the claims that come in, which must be submitted in writing, inquire into the validity of these claims, and single-handedly admit or reject these claims. Further, the officer is empowered to acquire land under the Land Acquisition Act of 1894, if this is necessary for the creation of the Reserved Forest.

Although Forest Settlement Officers had been appointed as part of the 1913 and 1915 notifications, a 1944 announcement listed a new Forest Settlement Officer for these areas,[55] suggesting that the previous officer had not actually completed his duties: claims had not been settled and compensations had not been awarded. Such confusions, both about the exact boundaries of the Reserved Forest and about the status of settlement efforts, have persisted up to the present day.

The application of the Indian Forest Act and the efforts to reforest the Ridge were just one component of New Delhi's creation. But there were other, broader trends in urban planning, both before and after Independence, that had a profound impact on the Ridge, and are thus crucial to understanding its current condition.

Planning the City, Planning the Ridge

From the beginning, one of the motives driving the design of New Delhi was a fear of density, and especially of densely packed "native" populations. New Delhi was consciously set up in contrast to the (supposedly) overcrowded, unsanitary, riotous lanes of Shahjahanabad, which, through this juxtaposition, became "Old" Delhi. But the very construction of the new city would inevitably bring a mass of Indians into Delhi, as builders, miners and other workers. To contain this flow of migrants, and to house people who would be displaced by the new city, the British planned to build the "Western Extension", so named because it fell to the west of Shahjahanabad, in an area now known as Karol Bagh.

Much of this extension was located on and around the Ridge, an area useful to town planners because of its relatively low population density (a result of the Ridge's traditional ecological function as a pastoral, rather than agricultural zone). British officials had no compunction about leveling parts of the Ridge to build housing, even while it beautified other parts of the Ridge and lauded their magnificent views. But once the leveling was complete, the actual execution of the Western Extension project faltered. The government did not build enough low-income housing or create sufficient urban infrastructure; instead, relatively wealthy Indians from surrounding areas built their own houses and took control of the area.

In response to the failure of the Western Extension and other schemes to relieve congestion and improve sanitation, British officials created the Delhi Improvement Trust (DIT) in 1937. The Trust, dominated by nominated appointees rather than elected officials, was tasked with clearing slums, building infrastructure and widening roads. Its most important responsibility was developing new planned neighborhoods, which could house up to 200,000 residents of the city. The government gave the DIT control over the huge tracts of land it had acquired for building the city, with the expectation that the DIT would subsidize the cost of developing neighborhoods by leasing the land at profitable rates to the new residents. From the beginning then, even though urban growth in New Delhi was state-dominated, it was largely driven by a logic of profit-making.[56]

The state monopoly on urban land, combined with the DIT's profit-driven approach, led to a uniquely dysfunctional land market in Delhi, which was exacerbated by the rush of Indian and international soldiers and military officers to Delhi during World War II. In this setting, the DIT was widely accused of profiteering and land speculation. Instead of building enough housing, the DIT instead built up very limited plots and sold them to the highest bidder (the "war-rich", as one commentator dubbed them). By encouraging scarcity in the market, the DIT pushed up land prices, and then profited by increasing rents. They also used the Land Acquisition Act to buy up land at cheap prices, and then sell it on the open market once prices shot up. This dynamic made it extremely difficult for the poor to find adequate housing.

Such conditions of scarcity and concentrated state power lent themselves to corruption. Government officials colluded with contractors to build houses made from black market materials that had been rationed for war efforts, and the houses were then rented out at exorbitant prices. The corruption was extremely widespread, affecting the DIT itself, as well as the police, customs officials, and the Public Works Department.

After Independence, the new leaders of India inherited the DIT, which was folded into the Ministry of Health. But before they could turn to reforming the DIT, which most Indian politicians resented greatly, they had bigger problems to deal with. The huge influx of Partition refugees put an enormous strain on Delhi's urban infrastructure, as well as its social fabric.

As Gandhi fasted in protest against the slaughter of Muslims and the vandalism of Qutb Sahib's shrine, other Congress leaders took a more pragmatic approach, and sought to find ways to safeguard the order and aesthetics of their seat of power. In addition to capping the number of refugees allowed in the city, the government set about building refugee colonies in areas then seen as far-flung outskirts of Delhi; once again, the Ridge's relatively low population density made it a favorite spot for government land-development projects, and refugee settlements sprung up in distant Ridge areas like Kalkaji. But roughly two-thirds of the refugees officially allowed in Delhi ended up, not in these resettlement colonies, but in properties owned by Muslims who had been forced to flee the city, including in Ridge areas like Mehrauli. Further, many who were not lucky enough to be included in the city's refugee quota stayed in the city anyway, squatting on government land that was controlled, but had not been developed, by the DIT.

After Independence, a growing population (including many non-refugee migrants), combined with lack of infrastructure, led to growing health threats, which culminated in a deadly cholera epidemic in 1956. This prompted Raj Kumari Amrit Kaur, Minister for Health and Local Self-Government, to write to the head of the Ford Foundation's Delhi office, and to ask for help planning for the future development of the city. The American foundation, whose initial money came from Henry Ford's industrial wealth, was becoming increasingly involved with philanthropic activities overseas, and readily agreed to the proposal. They even secured the expertise of Albert Mayer, a famous town planner from the United States.[57]

Mayer and his Ford Foundation team collaborated with the Town Planning Organization (TPO) established by the Indian government in 1955 to deal with the spiraling growth of the city. Shortly after Mayer arrived in India, the government also formed the Delhi Development Authority (DDA), which would carry out the plans that Mayer and his colleagues developed. The DDA was supposed to be a more sensitive, social justice-oriented version of the DIT, which it replaced. Even at the time, however, many opposition leaders claimed that the DDA, given monopoly power over land development in the city, would simply repeat the mistakes of its predecessor.[58]

There were other notes of discord as well. The (Indian) TPO team had an often-tense relationship with the (American) Ford Foundation team, as the former resented the latter's breezy paternalism. Just as British officials like Beadon claimed they knew what was best for the Indian populations, even as they consistently misread the situation on the ground, the American planners exuded a mix of confidence and disdain. It hardly helped that, besides Mayer, none of the American planners had significant experience with Asian cities, let alone Indian ones.

However, the clash between the American and Indian planners was largely one of egos, not ideology; most of the TPO planners had studied urban design in the United States, and they shared Mayer's fundamental beliefs about city planning. This ideology was its own kind of paternalism, with an emphasis on top-down planning that would create a rational, well-ordered city to replace the haphazard one built by waves of migrants. For the next six years, the two teams worked together to create a comprehensive planning document called, in its final form, the 1962 Master Plan for Delhi.[59]

It was no coincidence that Americans and Indians were brought together in this city, at this time. The Ridge, and the city as a whole, was once again drawn into much larger geopolitical currents. If, at certain critical points in British history (in 1857 and again in 1911), the Ridge took on an outsize role in imperial imagination, then the assumption of power by the newly independent Indian government on Raisina Hill signaled a different, radically new international alignment. This was characterized by the break-up of European empires after two world wars and long anti-colonial struggles, the corresponding emergence of independent governments in the former colonies, the rise of the United States as the world's most powerful nation (replacing Britain) and the onset of the Cold War.

One of the United States' chief geopolitical priorities was to keep countries in what was then known as the "Third World" from joining the "Second World", that is, the sphere of international Communism. If they had to, they would go to war to do this, as the conflicts in Korea and Vietnam later demonstrated. But there were other, less militant, responses. The Ford Foundation, for instance, saw itself as "moderating the cruder elements of anti-Red hysteria in the USA in the 1950s".[60] It prided itself on a more liberal internationalism, one which promised "Third World" audiences, especially elite postcolonial leaders, a way to address poverty and unrest in their countries while avoiding the tumult and upheaval of revolution.

In keeping with this model, the Master Plan was an attempt to make Delhi a "modern" city, one that could be a model for India and for the "Third World" as a whole. The Master Plan was part of a much larger Western focus on India; the country was the largest recipient of Ford Foundation funding in the 1950s.

One strategy for creating a modern Delhi was through zoning or restricting certain areas of the city to certain uses: residential, industrial, commercial, recreational, and so on. This was largely an extension of British methods of state control through mapping, drawing boundaries, and regulating activities like grazing. Another, much more radical strategy was the attempt to create a new urban citizen, unburdened by the fetters of tradition. One way of doing this was through the creation of open, green spaces like parks and gardens, where people from all walks of life could congregate, and where a new model of public life could emerge. Although these would ostensibly be democratic spaces, the imagined "model citizen" was generally identified, implicitly or explicitly, with the urbane, cosmopolitan elite. This idea was not unique to Delhi; when designing perhaps the world's most iconic urban green space, Central Park in New York City, Frederick Law Olmsted imagined it as a place where the working class could come and observe the manners and social graces of the upper classes.

More generally, the parks were meant to counteract the congestion of city life, giving residents an open space to get clean air. Mayer saw himself as part of the Garden City tradition which, among other things, decried the deleterious effects of living in dense, polluted cities, and advocated for more dispersed growth that combined the best of the rural (open space, healthy living) with the best of the urban (employment opportunities, vibrant social opportunities). Urban parks were a central part of this vision, bringing the benefits of village life into the city.

Ironically, this emphasis on the restorative impact of green space coexisted with concerted attempts to banish any trace of the rural from Delhi. In many planning documents from this time, the rural is conflated with the backward and the uncivilized. As part of its zoning regulations, the Master Plan prohibits "village-like trades" in the urbanized part of Delhi, on the grounds that they "cast an unhealthy influence on the urban setting". For the planners, "milch cattle and dairymen" were of particular concern. The Master Plan gives provisions to relocate them to "urban villages", pockets of rurality exempt from the plan's strict regulations. These were generally pre-existing villages that were swallowed up by Delhi's rapid expansion. As the scholar Ravi Sundaram notes, "This was in effect a program for a modernized form of urban apartheid, a purging of the village from the city through legal banishment to designated enclosures."[61]

Another way of enforcing this apartheid was through the promotion of a strictly rural, agricultural "Green Belt" that would encircle the urbanized core of Delhi but would not take part in any urban activities. This agricultural belt would, in theory, help feed the city, while also containing urban sprawl.

The ambiguous attitude towards green areas in Delhi (praised as parks; marginalized as productive landscapes) led to further transformations on the Ridge. One the one hand, the Ridge was imagined largely as a recreational space; in the 1962 Master Plan, it is referred to several times as a "Regional Park". The model is clearly Western. Commenting on the Ridge just north of Raisina Hill, the Master Plan states, "It should not be allowed to be dissipated by small undesirable uses but should be gradually developed as a central public park in

Delhi, comparable to Hyde Park in New York."[62] The transformation of the Ridge into a park was envisioned as a slow process: "small portions of it may be developed... and the rest may remain in a natural state with the under growth."[63] The difficulty of keeping any of Delhi's land in a "natural state" would soon become evident.

While parks were promoted, the long-suffering pastoralists of the Ridge were dealt yet another blow. Their profession dismissed as non-urban and hence backward, they could either submit to state displacement, find another career or continue keeping livestock on the sly. Many chose the last option. The Master Plan was notoriously difficult to implement, since it called for a total reordering of urban life, in a city where "illegal" settlements (especially on DIT land) had already proliferated and become a fact of life. Many government officials, especially the ones working on the ground, also had a material interest in bending the rules of the plan. They could collect bribes on a regular basis and build up networks of power and patronage.

Such dynamics continue to this day. In the Gujjar-dominated village of Chandrawal near the northern Ridge, which suffered under both Mughal and British rule, the marginalization of their pastoral livelihood continues. This iteration of Chandrawal is now sandwiched between two neighborhoods planned in the 1950s to accommodate Partition refugees, Jawahar Nagar and Kamla Nagar. Chandrawal is classified as an "urban village" so the residents are technically allowed to keep cattle within the bounds of the neighborhood. But there is little space to graze, so the cows are let out to roam around the surrounding neighborhoods, eating trash and blocking traffic. Many residents also run dairies that have not been authorized by the government. Both these activities are, technically, illegal, but the residents know who to pay off. Further, as the scholar Amita Baviskar notes, "The fact that Chandraval village also has a sidebusiness in supplying musclemen for local politicians, property brokers and others who need someone to lean on or be leant upon also helps give its dairies a degree of immunity from harassment by city authorities."[64]

The ongoing resistance of Chandrawal villagers, and their uneasy alliance with a range of officials, is indicative of the larger failure of the Master Plan. The grand ambitions of the Master Plan quickly dissolved, overwhelmed by the realities of urban life and governance in Delhi. Beyond the impracticality of the new zoning regulations, there were massive problems with land development, as the DDA followed in the footsteps of the DIT. The basic issue was the same: the DDA, like the DIT, still refused to build enough housing, instead developing small plots and selling them at relatively high prices. The dearth of low-income housing was especially pronounced; the growth of informal working-class settlements (pejoratively dismissed as "slums") was largely in response to this failure.

The continuity from DIT to DDA indicates the similarities between the British and Indian approaches to city planning, not least in their emphasis on a heavy state hand, and their prioritizing of the urban at the expense of the rural. Both these tendencies were evident in their respective land acquisition strategies, which used the same legal tool: the Land Acquisition Act of 1894.

When building New Delhi, the British acquired roughly 35,000 acres of land, largely from the surrounding villages, including several on the Ridge. The DDA outdid them, acquiring roughly 39,500 acres in 1959, in the biggest land nationalization outside of the Communist world. Again, this was done at the expense of Delhi's villages, as land prices were once more frozen by the initial notification under the Land Acquisition Act, and meager compensation was given to villagers whose farmlands and village commons were purchased by the state. Due to the Master Plan's "urban village" provision, however, villages were not totally uprooted, unlike the cases of Malcha and Raisina; instead, the residential core of villages (called the *abadi*) were allowed to stay, while the villagers' means of livelihood (farming land, grazing land, wood lots, etc.) were snatched away.

This enormous land-buying spree is sometimes referred to as the "Delhi Experiment". It is universally recognized as a failure. The 1962 Master Plan had the goal of planning and controlling Delhi's growth for next twenty years. Halfway into this time period, it was abundantly clear, even to the government, that the plan had been a bust. A report written by the TPO's successor, the TCPO (Town and Country Planning Organization), cataloged the following failures: near-universal class-based segregation; extensive violations of the meticulously planned zoning rules; a lack of housing options, especially for the poor; and widespread squatting on DDA land. An earlier report bemoaned the skyrocketing of land prices, which were largely the result of DDA-driven land speculation; land prices had jumped nearly four times from 1958 to 1969.[65]

On the Ridge, these failures were particularly clear. Historically, because of its rocky soil, the Ridge has been sparsely populated. It thus became a popular spot for "illegal" housing of all kinds, from super-elite mansions to desperately poor settlements. As we shall see in the next chapter, the ultra-rich took advantage of the supposedly "rural" character of the far reaches of the Southern Ridge—part of the Master Plan-mandated "Green Belt"—to build lavish homes they dubbed "farmhouses" to keep up the legal pretext of their settlements.

A New Beadon

One of the young officials who witnessed the disintegration of the Master Plan's grand hopes was Jagmohan Malhotra, usually referred to simply by his first name. Jagmohan was outraged at the way the Master Plan's norms were being violated, and much of his long, accomplished and, to many, notorious career can be seen as an attempt to create the ordered city that the Master Plan had promised.

Though Jagmohan was inspired by the American-influenced Master Plan (which he called "a pioneering work") he bore more of a resemblance to European figures, and especially to the British officials that once ruled Delhi. Jagmohan was, in many ways, the Beadon of post-Independence Delhi. He is not always mentioned in mainstream accounts of the city (just as everyone knows Lutyens, but few know Beadon), but if one dusts the surface just a bit, one finds his fingerprints everywhere. His influence on the city as a whole is remarkable; his particular interventions on the Ridge are also surprisingly pronounced.

Like Beadon, he presents himself as a hard-headed but ultimately kind-hearted administrator, making the tough decisions that may attract enemies in the short term, but will benefit society in the long term. In one of his writings, he registers his agreement with "Lord Wavell's observation: 'India can be governed firmly or not at all.'"[66]

Throughout his long career in Delhi, he remained convinced that the problem with the urban planning in Delhi was not the DDA's total inability to build adequate housing for the poor, nor the perverse incentives the DDA's land monopoly gave for corruptions large and small. No: the problem was "the soft and permissive nature of the Indian state".[67] Government officials simply lacked the will-power to stand up to builders, real estate developers, and "encroachers", and sternly implement the letter of the law, for the greater good of society (as he saw it).

He had the opportunity to combat this chaos relatively early in his career. As an Implementation Commissioner for the DDA in the 1960s, he saw myriad failures of the Master Plan. By 1975, he had risen to the position of Vice-Chairman of the DDA. In June of that year, when Prime Minister Indira Gandhi imposed a State of Emergency on the nation and began ruling by decree, Jagmohan found his opportunity to re-order Delhi in the ways he had always dreamed of doing.

Jagmohan had long been an admirer of Baron Haussman, the nineteenth-century figure who had overseen the renovation and modernization of Paris, giving the city wider, more "rational" roads and uprooting many traditionally working-class neighborhoods. Most of the changes were made in the name of order and control, especially given Paris's history of mass demonstrations; the volatile working classes would be both more scattered and more easily controlled by police stationed on the spacious new roads. The creation of green areas was a major part of the reconfiguring of Paris, and many buildings were razed to make room for parks. One of Jagmohan's chief aims was to reshape Delhi in much the same way Haussman had reshaped Paris.[68]

He found willing partners in Indira Gandhi and her son Sanjay, who oversaw most of the more unsavory aspects of Emergency. Many in Delhi were targeted during the brutal "cleansing" of the city during Emergency (both Indira Gandhi and Jagmohan were fond of the "cleaning" metaphor), including left-wing students in Delhi's universities and colleges, opposition politicians from both the left and the right, and intransigent journalists (of whom there were alarmingly few). But the poor undoubtedly bore the brunt of the Emergency's disciplinary force. Jagmohan, like many British and Indian officials before him, had an instinctive revulsion to crowded, low-income areas, especially in the Muslim-dominated old city of Shahjahanabad. In a book he wrote a year before the Emergency, Jagmohan had this to say:

> *Certain parts of Shahjehanabad have become dead—infrastructurally and culturally. In Hauz Qazi, Lal Kuan and Turkman Gate, bums and bad characters are all that can be seen at nightfall. Only indecent remarks of cheap film songs are heard. The very sight of eating places with broken chairs, stinky tables with shabby and shirtless waiters, is repellent. It is necessary to brighten up these areas; otherwise these will remain the breeding grounds for criminals and rioters.*[69]

The Emergency targeted such areas, even though the poor were hardly the only ones breaking the norms of the Master Plan; they simply made a more convenient target for the state's force. About 60,000 families were displaced, their homes demolished by government bulldozers.

Jagmohan was widely reviled for his role in the Emergency. He was stung enough by this criticism that he wrote an entire book defending himself, called *Island of Truth*.[70] Jagmohan justified the demolitions on largely aesthetic terms, in words that resonate with the British creators of New Delhi, and all the state actors that have sought to portray the glory of the capital. In a later book, he writes,

> *After the execution of the clearance-cum-resettlement-cum-redevelopment project, Delhi looked a neat, clean, orderly and organised city with a personality and identity of its own. Areas around historical monuments were cleared, landscaped and developed as parks and community greens. Not only was the architectural and cultural legacy of the metropolis preserved but also airy lungs were created all around.*[71]

Jagmohan's celebration of greenery combined the British emphasis on grandeur with the American emphasis on parks as a means of moral improvement, to transform the "bums and bad characters" of Delhi.

Several years after the Emergency, Jagmohan was .... to the position of Lieutenant Governor of Delhi, the highe....... .tive position in what was, at the time, officially known a...... Union Territory of Delhi. This means there was no state legisla....... .nother attempt by the central government to keep control over its capital— and that the Lieutenant Governor had considerable power, as long as he did not run afoul of the leaders on Raisina Hill. By this time, the anti-Emergency backlash had ended. Indira Gandhi became Prime Minister again in January 1980, and Jagmohan was named Lieutenant Governor of Delhi in February.

In this role, Jagmohan oversaw the biggest expansion of Delhi's protected forests in decades. After the 1913 and 1915 notifications of the Ridge (Central and Northern, respectively) as Reserved Forest, subsequent legal notifications had simply tinkered with their boundaries, or appointed new Forest Settlement Officers to take up the seemingly-endless task of settling claims to compensation for forest land. This was true before Independence and after; a 1958 notification named yet another Forest Settlement Officer, and a 1965 notification redrew the boundaries of the Central Ridge.[72]

Jagmohan changed this trend. He was the first in post-Independence Delhi to use the Indian Forest Act to actually notify sizable plots of new forest land. Always ambitious, he notified more than twice the amount of forest land than his British predecessors had; 5,853.9 acres to their 2,343. He had the land notified as a "Protected Forest", which is slightly less stringent than a "Reserved Forest", and allows for more local uses, though still under the state's watchful, punitive eye. The notified land was composed of 25 sites scattered across Delhi, with the majority of the land on various parts of the Ridge.[73]

The seeming precision of the figures involved in the notification concealed a much messier reality on the ground. According to the official notification, all of the land involved already belonged to the government; it had either been acquired by the DDA as part of the "Delhi Experiment" or was owned by the Land and Development Office (L&DO), a colonial-era institution that had handled much of the acquisition and land record administration during the building of New Delhi.

However, these acquisitions, as we have seen, had been quite contentious. They were resisted by the villagers who were losing their farmland and their commons, and later, many DDA and L&DO plots were occupied by those who migrated to Delhi. Sometimes, these two groups formed uneasy dependencies. Some villagers, for instance, rented out land that used to be their village commons, but was now technically government owned; migrants had little choice but to accept these arrangements, which often involved bribes to local police and other officials. Jagmohan's forest notifications did little to change this tangled web of relations.

Further, there was still a lingering sense of public resentment and outrage at the large-scale housing demolitions of the Emergency, so Jagmohan did not have the leverage to push through another round of demolitions to clear the way for parks and forests. This agenda would have to wait for another decade or so, when the public mood had changed yet again, as we'll see later in this chapter.

As Lieutenant Governor, Jagmohan still found ways to reshape the city, though his actions suggest that his emphasis was not primarily on ecology, but on the larger message that greenery could send. The greenery itself was expendable, if there were other, more powerful ways to send that same message. This was evident in Jagmohan's central role in planning the 1982 Asian Games, a kind of regional Olympics, with athletes from 33 Asian countries competing. Hosted in Delhi, the Games were, in the eyes of the organizers, including Indira Gandhi, yet another way to equate the city with larger glory and grandeur. Jagmohan writes that the Games "won applause from thousands of visitors" and "enhanced India's prestige abroad" thus giving "new self-confidence to the nation".[74] Live telecasts of the Games, a novel phenomenon in India at the time, spread the event around the country instantly, and strengthened the symbolic connection between Delhi and India as a whole.

On a more material level, the Games involved a major reordering of the city, including the building of eight new stadiums and a "Games Village" to house the foreign athletes and coaches. The Games were also used as an opportunity to push through several infrastructure projects, including a local railway line circling the city. To handle this construction boom, contractors and builders brought in huge numbers of migrant workers. One account estimates that roughly one million workers migrated to the city in the lead-up to the Games.[75]

More local sources were tapped as well, both for labor and for raw materials. Badarpur sand from Bhatti Mines and other Ridge sites was an essential ingredient for many construction projects, and quarrying operations kicked into high gear to meet the increased demand. But this was not the only way that the Bhatti Mines area factored into the Games preparations. The government decided that one of the new athletic facilities, a shooting range, would be located on a patch of land adjacent to the mines, which quarry workers had long been using as a grazing ground for their livestock and as a playground for their children. The government had little regard for such uses of the land, and construction of the shooting range soon began.

The government's attitude to the workers, and especially to their grazing activities, is, perhaps, to be expected. More surprising was the government's stance on a patch of forest land that Jagmohan had seemed to intend to promote. The land in question was in the vicinity of Siri Fort, the ramparts of the walled city built in the early fourteenth century by Alauddin Khilji as a defense against Mongol invasions. It served that purpose well, but now, like all of Delhi's old cities, it lies in ruins. Before 1982, these ruins were surrounded by forest. In order to build the Asian Games Village, much of the forest was cleared, to the alarm of the nascent environmental movement in the city.[76]

It is unclear how much glory, and how much international attention, the Games actually brought to Delhi. Despite Jagmohan's references to thousands of visitors, the overwhelming majority of foreigners who came to Delhi were either athletes or officials. Only an estimated 200 foreign tourists and spectators came for the event. Further, the construction for the Games was extremely costly, and the costs were hardly recovered. Many of the stadiums lay vacant after the event, unable to create revenue, until they were renovated in another costly procedure for another costly mega-event: the 2010 Commonwealth Games. An athletes' dormitory called the Players' Building was not finished by the time the 1982 Games started and remained half-built for the next 15 years. The influx of workers exacerbated Delhi's chronic housing crunch, leading to even more of the informal settlements that Jagmohan so despised.

Jagmohan still portrayed the Asian Games as a victory, since it was, after all, a successful media event, seen by all the country. This was Jagmohan's last major attempt to reconfigure Delhi. From 1984 to 1990, he served as the Governor of Jammu and Kashmir. His stint as Governor, during a raging separatist insurgency, was marked by controversy. He was sent by the government because of his reputation as a firm administrator, but, as one journalist put it, he used his post to usher in "a period of unfettered repression".[77] As usual, Jagmohan defended himself by writing a book, entitled *My Frozen Turbulence*.[78]

By 1996, Jagmohan had left Congress and joined the BJP, and he was named Minister of Urban Development in 1999. This gave him an opportunity to revisit Delhi, which he describes in terms that are quite literally paternalistic.

I was deeply agonised. In my earlier assignments in the DDA and as Lieutenant Governor, I had nursed Delhi like my child and ensured its planned development on the ground, strong opposition from vested interests notwithstanding. I found that this child had been badly bruised by all and sundry.[79]

Unable to beat back the collective force of these "vested interests" and resented by his colleagues for the friction he was causing, Jagmohan was shifted to the less powerful position of Minister of Tourism. Excluded from the large-scale modernist urban restructuring projects that he so admired, he instead took up smaller interventions. Several of these involved the use of parks to convey the majesty of the state by invoking past dynasties. Given his increasing alignment with the Hindu nationalist rhetoric of the BJP, he could hardly invoke the Mughals or Sultans; in proud Hindutva tradition, he chose to glorify Prithviraj Chauhan as the savior of Delhi. Beyond the larger philosophical and political problems with his reading of history, it also simply gets the facts wrong. Although Prithviraj controlled the Delhi region, he never used it as a capital; as mentioned earlier, he was based in Ajmer.

This did not stop Jagmohan from installing a huge statue of Prithviraj Chauhan in the Southern Ridge, where the ruins of the old Rajput fort of Lal Kot meet the ruins of Jahanpanah, the walled city built by Mohammad bin Tughlaq. The statue was inaugurated and unveiled by the Home Minister, L. K. Advani, better known for his role in the demolition of the Babri Masjid. Several months later, Advani led a prayer ceremony at the nearby Qutb Minar, in an attempt to reclaim it as a Hindu space.

Advani's view of history may be skewed and reductive, but he, like Jagmohan, understands the role of symbolic spaces in projecting state power. Although he might deny the legitimacy of the sultans who built the Qutb Minar and the surrounding city, he, whether consciously or unconsciously, picked up a crucial lesson from them: that states rule not just through military might, but through metaphors, symbols, and religious and political spectacles, whether they are architectural splendors, sporting mega-events or prayer services. Advani could not deny the significance of the Ridge-side minaret, Delhi's abiding symbol of state power, even as he tried to imbue it with new meaning.

PART III: SERVICES

The Ridge, then, has been the site of both pitched military battles and symbolic demonstrations of the state's legitimacy, grandeur and righteousness. But the state cannot sustain itself only through shows of force and strength, whether literal or symbolic. Beyond coercion and propaganda, the state also needs to provide at least some basic services to at least some of the population. This is not divorced from the use of military power; as the state expands through military might, it also promises (though does not necessarily deliver) stability and peace to its inhabitants. Of course, the scope and range of services provided (and expected) varies dramatically through time and space, with the modern era witnessing a huge proliferation of state programs implemented for the benefit of citizens, from health care to education to employment schemes.

But the services a state provides are not necessarily an unalloyed good. They benefit some more than others, and can, in many cases,

harden existing power relations rather than challenge them. What is more, the idea of "public services" raises crucial questions: who is considered part of the public? Are slaves? Are women? Are the so-called "low" castes? Are foreigners?

Deforestation for the Public Good

These general questions have had very specific resonances in the exercise of state power in Delhi, and on the Ridge in particular. And these questions go back to the early days of the Sultanate, when Delhi was first being established as a political capital. In those times, modern terminology like "public services" was hardly current; a more religious idiom was used. But just as Qutbuddin Aibak and his rival Bahauddin Tughril used authoritative religious terminology to make very political and very contested claims about their sovereignty over North India, early sultans like Shamsuddin Iltutmish invoked religion to give symbolic weight to the mundane tasks of governance on the edge of the Arid Zone.

In 1230 CE, as the legend goes, Iltutmish had a dream. The Prophet Muhammad appeared before him, riding his horse through the hilly terrain at the outskirts of Iltutmish's newly powerful capital city of Delhi. The horse stopped in the middle of a low basin and tapped his hoof on the ground. In the dream, Iltutmish understood that it was his sacred duty to build a reservoir at this spot. Several days later, Iltutmish was surveying the land around his city, and he suddenly stopped, recognizing the landscape from his dream. He approached the spot where the horse had halted, and he saw a lone hoof print in the dirt. He immediately ordered that a water tank (or *hauz*) be dug at this spot.

The tank took on his name and became known as Hauz-i-Shamsi. It was a grand construction, with a pavilion exactly in the center, where the hoof print had been found, to commemorate the prophetic vision. It became a crucial water source for his city, and domestic servants would make the trip from the city center to the *hauz* on the outskirts to collect water and do laundry.

It is not surprising that the *hauz* was freighted with such symbolic and religious significance. Water was a precious resource, not easy to find in the Arid Zone. Therefore, providing his people with water was one of the most significant services a ruler could provide. The Ridge, with its variable slopes and its rocky soil, was a fickle source of water. However, if one dug at the right place, one would be rewarded with plentiful groundwater that had been filtered through the porous quartzite of the Ridge and was hence remarkably clean.

But the Hauz-i-Shamsi suffered from the political turmoil that enveloped Delhi after Iltutmish's death. Different groups of powerful slaves propped up a series of short-lived sultans, until, in 1266, a Turkish slave named Ghiyasuddin Balban was finally able to wrest control of the empire from his rivals. He made many efforts to consolidate the Sultanate's power and to expand his reach farther into the hinterland of northern India. But one of the first problems confronting him was more immediate. Hauz-i-Shamsi, and the area around it, was plagued by attacks and thefts. Domestic servants were beaten up and robbed when they went to the *hauz* to do their daily chores, and people from the city did not feel safe traveling beyond the walled city at night, for fear of attacks.

The assailants were from the Meo community, a group that has much in common with the Gujjars. Both, until the colonial era, largely relied on a nomadic, pastoral lifestyle, and both were typically portrayed as dangerous, uncultured and unreliable by those living in more settled communities. But both were also admired for their rebelliousness and political independence.

Balban had little patience for the Meos. He realized that they were hiding in the forests around the *hauz* and in the outskirts of the city more generally. One of his first acts as Sultan was to clear the trees at the edge of the city, so the Meos would have no place to stage their attacks. He then launched an aggressive military campaign against the Meos, bringing the region surrounding Delhi firmly under his control. He used this same method to expand his empire, hiring specialized woodcutters to clear paths through forested areas, then relocating intransigent populations, enslaving women and children, and promoting agriculture by offering the newly cleared land to loyalists.[80]

Both the deforestation and the military action could be considered a public service; certainly, for the domestic servants who had to make their daily trips to the *hauz*, life was easier. But the "public" had strict boundaries; for the nomadic and pastoral communities of the area, Balban's reign was disastrous.

Forestation for the Public Good

This was hardly the last time the state used deforestation as a military strategy. After their victory over the rebels in 1857, British officials went on a tree-cutting spree, deforesting large swaths of land between the Ridge and Shahjahanabad, and in the city itself, both as punishment and as a way to facilitate easier surveillance of the population.

But, as we have seen, British interest in the Ridge eventually turned to afforestation, especially once the construction of New Delhi began. And even though the British rationale for afforesting the Ridge was largely aesthetic and political, environmental concerns, and their impact on humans, featured as a secondary impetus. One planning document plainly noted that, while aesthetic factors were the most important, afforesting the Ridge would also have the effect of stopping the run-off of rainwater, and hence preventing soil erosion.[81]

For Jagmohan, the emphasis on parks and forests was clearly part of a larger aesthetic vision of a "clean and green" city that would promote Delhi's prestige both locally and internationally. But Jagmohan, even more than the British, described parks and greenery as a public good, something that would benefit the residents of the city. Using a bodily metaphor that has become increasingly popular, Jagmohan emphasized the need for a city to have lungs so that it could breathe easily and cleanly.

During Jagmohan's time as Lieutenant Governor, another kind of environmental rhetoric was emerging in the city, one that emphasized wildness and transcendence over a tidy, state-imposed neatness. This was in sync with the Romantic strains of an environmental movement that was gaining global recognition. In terms of Ridge activism, the pioneers were a group of students who eventually founded a group called Kalpavriksh (named after the bountiful tree of Indian myth). They were moved by the messier parts of the Ridge, the untamed patches full of thickets and undergrowth and wildlife. As a later report on the Ridge opined, "unless one is exposed to its rough terrain, its adventure, smells and sounds, it is difficult to sense the peace it can provide."[82] Their engagement with environmental activism was often sparked by deeply emotional connections to the little-explored eco-systems of the Ridge, and one of their first organized activities was leading nature walks with other college students.[83]

Their vision of the Ridge stood in marked contrast to the actions of the state post-Independence, and to the idea of the Ridge as a prim, proper, manicured space. The Master Plan had envisioned the Ridge as a Regional Park, and later administrations were happy to comply with this vision. In 1962, the Central Ridge around Raisina Hill was handed over to the Central Public Works Department, so that it could be developed and beautified as a set of parks; in 1968, a similar order, with similar rationale, transferred the Northern Ridge to the DDA. Further, in the decades after Independence, the government had done little to maintain the larger Ridge as a forest.

The student-activists of Kalpavriksh were keenly aware of this. In 1979, one of the students found out that the government had allotted a plot of land in the Central Ridge for building schools. They organized a protest rally, gathering support from the residents of the neighborhood as well. Though they were unable to stop the construction of the schools, they had laid the foundations for Ridge-centric environmental activism.[84]

Despite their criticism of state attitudes and actions towards the Ridge, the students found that some very high-level government figures were receptive to their demands. In 1980, the students sent a petition to Indira Gandhi asking that no new constructions come up on forest land. Gandhi's urban vision was quite well aligned with Jagmohan's (as the Emergency and the Asian Games had shown). She also shared his environmental sensibility; in the late 1970s, she had requested the Indian Air Force to modify a planned construction on the Ridge so that the city's skyline and the forest could be preserved (note the twinning of aesthetic and environmental concerns). She was also responsive to the students' petition, which likely contributed to Jagmohan's notification of 25 new forest zones.

However, these limited steps could not counteract Delhi's explosive population growth, which necessitated new "illegal" settlements for the poor and the rich alike. So while environmental consciousness grew, it did little to change the larger social, political and economic dynamics that were eating away at the greenery of the Ridge.

Pollution and the Public Interest

The terrain of environmental activism in Delhi began to shift in the 1990s, largely due to a judiciary that was increasingly receptive to complaints about environmental issues, which were framed as matters of public good. A key figure in this development was the environmental lawyer M. C. Mehta, who filed a slew of legal petitions targeting pollution in the city. In his petitions, Mehta repeatedly invoked Delhi's Master Plan, arguing that its careful zoning rules and regulations, especially regarding industrial activity, needed to be strictly enforced.

Mehta's petitions are categorized as Public Interest Litigations (PILs), in which the complainant is not necessarily the aggrieved party, but rather a citizen or organization concerned with the public good. PILs proliferated in India after the Emergency, as the judiciary, which had been notoriously compliant during Indira Gandhi's experiment with authoritarianism, sought to re-establish its progressive credentials and prove that it was working on behalf of the downtrodden. PILs in the early 1980s thus largely focused on marginalized or oppressed groups. For example, in 1982, the People's Union for Democratic Rights (PUDR) filed a PIL on behalf of the construction workers who were employed building stadiums for the Asian Games and who were facing widespread labor abuses.

However, the PILs filed by M. C. Mehta were of a very different nature. They also claimed to be in the "public interest", but in a more nebulous way. Mehta argued that the residents of Delhi as a whole would have a higher quality of life if industrial pollution and other kinds of environmental ills, including air pollution and noise pollution, were reduced. Although Mehta filed scores of PILs addressing issues across India, one PIL in particular (to be precise, Writ Petition 4677 of 1985) would go on to have an outsize impact on the national capital.

Writ Petition 4677 has had a long life, working its way up to the Supreme Court, which issued a number of rulings related to the petition. The one that got the most attention was a 1996 ruling mandating that all "hazardous" industries should be relocated outside the city. This was later expanded to all "polluting" industries in "nonconforming" areas; that is, areas not zoned as industrial in the Master Plan.[85] In 2000, the government finally attempted to enforce this order, which rendered thousands of workers jobless. This led to a series of massive protests, which rocked the city for four days, as factory owners found a rare piece of common ground with their workers and encouraged workers to rally against the ruling.

This case suggests the ambiguities of "public interest". Lawyers like Mehta positioned workers' rights as simply a sectional interest, as opposed to environmental issues, which impact the entire population. But for the thousands who lost their jobs due to factory closures, marginally cleaner air hardly made up for the loss of livelihood. This brings us back to the realm of anti-ecological environmentalism. While the largest source of air pollution in Delhi is vehicular traffic, not industry, there has never been a Supreme Court order relocating all SUVs and luxury cars outside the city's borders. Moreover, this kind of environmental thinking draws on an aesthetic tradition, from Beadon to Jagmohan, that sees the landscape in terms of neatness, cleanness, and respectability, and does little to solve the root problems of ecological destruction. (Mehta, not surprisingly, was one of the chief proponents of banning mining in Delhi, a move which embodied precisely this kind of short-sighted logic.)

In the case of the Ridge, this anti-ecological environmentalism was only one of the streams of activism that emerged in the 1980s and 1990s. The members of Kalpavriksh, though also from an upper-middle-class background like Mehta, began to develop a much more thoroughgoing critique of consumer society and to question their own privileged role in that society. Nonetheless, their rhetoric, especially their emphasis on "encroachments" on the Ridge, could feed into a more narrowly conceived, top-down environmentalism.[86]

These different streams of environmentalism came together in 1992, when several portions of the Ridge were transferred to the DDA. Given the DDA's evident lack of concern for forestry, this move led to an outcry from Delhi's multifaceted environmental movement. Seventeen organizations banded together to form the "Joint NGO

Forum to Save the Delhi Ridge". The forum include   local groups like Kalpavriksh, as well as the Delhi branches of   .tional organizations like the World Wildlife Fund. The forum took to the streets with protests and marches, but it also used its economic and social capital to pursue other means of engagement. They enlisted well-known Delhi residents to write letters and sign petitions in defense of the Ridge, and they produced "Save the Ridge" commercials that aired during prime-time news hours. Noted author Khushwant Singh was on the advisory board of the forum. Their campaign was widely covered in the media, especially the English-language media, which saw itself as articulating the demands of a newly assertive middle class.[87]

Due to increasing public and media pressure, the government established a 10-member committee to give recommendations about the management of the Ridge. Led by Lovraj Kumar, an oil executive and civil servant, the committee included six government officials and four NGO leaders. The committee's report, released in 1993, was far-reaching, with 20 recommendations, but at its core was the demand that all of the Ridge be notified as a Reserved Forest, and that the boundaries of the forest be immediately demarcated. Other recommendations were officially calling the Ridge a Reserved Forest (as opposed to a Regional Park) in the Master Plan, minimizing the conversion of the Ridge into manicured gardens, and creating a Management Board to oversee the protection of the Ridge.[88]

One key recommendation dealt with the thorny issue of who should actually control the Ridge. After much debate, and with much trepidation, the committee recommended that the DDA should retain its control of the Ridge, with the exception of the Asola-Bhatti Wildlife Sanctuary, which was already under Forest Department control. This was deeply ironic, given that the "Save the Ridge" agitation had gained momentum precisely as an anti-DDA movement. However, the committee members reluctantly concluded that only the DDA had the power and the resources to properly manage the Ridge, a task that included demolishing encroachments, securing the Ridge's borders, and developing parks and schools outside the Ridge to relieve the pressure on Ridge land. The committee insisted, though, that the DDA should set up a separate Ridge Management Division within the agency, which would be sensitive to the unique requirements of forest maintenance as opposed to park creation.

The government did not follow through with all the committee's recommendations; the Ridge remains a Regional Park in the Master Plan, for instance. More alarmingly, the DDA never set up a separate Ridge Management Division. Even today, the Ridge is overseen mainly by the horticultural team at the DDA, to the dismay of many Delhi ecologists. The government did, however, agree to the committee's core demand; on 24 May 1994, the Lieutenant Governor of Delhi officially notified 7,777 hectares of the Ridge as Reserved Forest under the Indian Forest Act. This included the portions notified by the British in 1913 and 1915 (the Northern and Central Ridge), as well as areas that Jagmohan had notified. It also expanded many of these areas, especially in the Southern Ridge, where over 6,200 hectares, or more than two-thirds of the total land, was notified; this included the Wildlife Sanctuary that had swallowed up the erstwhile Bhatti Mines. The 1994 notification envisioned the Ridge as four discrete zones; the three already mentioned, plus the South Central Ridge, near the old Sultanate stomping grounds of Qutb Minar.

This notification was similar to Jagmohan's in that it did not, at first glance, require any additional acquisition of land or compensation to residents. The notification specifies that, in the four Ridge zones, Reserved Forests are to be created on "all forest lands and wastelands which is the property of the government, or over which government has proprietary rights."[89] Technically, then, this was simply a matter of transferring government land from one department to another. But as with the 1980 notification, the reality on the ground was much more complicated. A book about the Ridge issued by the Delhi Forest Department is quite frank about this, noting that the 1994 notification was issued "without a prior assessment of the availability of the Ridge land free from encumbrances".[90]

Further complicating matters, M. C. Mehta's Writ Petition 4677 reared its head again. In yet another response to this PIL, the Supreme Court, in 1995, alluded to the Lovraj Committee's recommendation and ordered that a Ridge Management Board be established. This board mirrored the composition of the Lovraj Committee; it was dominated by appointed government officials but included a few NGO members in deference to their role in Ridge activism. Because it was so stacked with government officials, it soon became a rubber stamp for government projects to build on the Ridge, although it was more successful in keeping private development out.

But M. C. Mehta and his unavoidable petition were to have an even more profound impact on Ridge management. In later hearings, Mehta argued that the 1994 notification did not actually extend to all of the Ridge, and specifically that it left out large parts of the Southern Ridge. In response, the Supreme Court issued two orders, one in January 1996 and one in March 1996, directing that "uncultivated surplus land of Gaon Sabha falling in the Ridge" should be "made available for the purpose of creation of Reserved Forest".[91]

These cryptic orders need some decoding. They rely, not on the Indian Forest Act, but on the Delhi Land Reforms Act, a piece of legislation that was meant to be a progressive contribution to rural Delhi. Among other provisions, the Act transfers control of village commons from the *maliken deh*, or traditional ruling village body made up only of (male) landowners, to the *gaon sabha*, which includes all adults in a village. However, the Act also contains a provision that calls to mind old British land settlement policies; it mandates that "if the uncultivated area situated in any Gaon Sabha area is, in the opinion of the Chief Commissioner, more than the ordinary requirements of the Gaon Sabha, he may exclude any portion of the uncultivated area from vesting in the Gaon Sabha." In essence, if a government representative decides a village has too much land, he can take that land away from the village, with no compensation required.

The law also reflects old prejudices. Farmland is protected, but "uncultivated" commons are viewed, essentially, as wastelands, with all their pastoral uses rendered invisible. This was bad news for villages on the Ridge; given its rocky nature, the great majority of *gaon sabha* land in these villages was uncultivated, and hence became an easy target for the state. Much of this land had also been used for quarrying. Although this had been officially banned in 1991, it still continued surreptitiously in some places. This further drew Mehta's ire; here was another example of a "polluting industry" failing to conform to state mandates. The Supreme Court concurred.

In response to the Supreme Court orders, the executive branch acted quickly. In April 1996, Delhi's Lieutenant Governor issued another notification, instructing 14 villages on the Southern Ridge to turn over their "surplus" common lands, for the public good, of course. But this was no easy task; given centuries of varied land uses and changing legal systems, it was difficult to determine what land, exactly, formed part of the village commons, and which belonged to individuals. Further, it was far from clear what counted as surplus.

It did not help matters that the state itself sometimes seemed confused about these distinctions. In 1994, the government acquired land from villagers in Tughlaqabad, a Gujjar-dominated settlement set amidst the ruins of Ghiyasuddin Tughlaq's old city. The fact that this acquisition even happened sheds doubt on the claim that all the land notified in 1994 already belonged to the government. Several of the villagers were unhappy with the compensation amount and filed a lawsuit; the High Court sided with the villagers and ordered that they be given a much higher compensation, namely ₹132 crore. The government appealed this decision, but the Supreme Court upheld it.

By the time the government began the paperwork for paying the compensation, the 1996 notification had been passed, and the government introduced a new claim: that the land in question was actually uncultivated village commons on the Ridge, and thus no compensation was necessary. The government filed a new case, which the court has yet again rejected; but the government has appealed yet again. These proceedings reflect the ambiguous status of Gujjars in Delhi; they are powerful and organized enough to fight a continuing legal battle with the government and to win most of their cases, but they do not have enough clout to get the government to actually pay up, yet.

For other, less locally dominant groups, the fight has been even more difficult. One court case involves a petitioner who said he and his family had been allocated a plot of land under Indira Gandhi's 20-point program, a government scheme that included half-hearted land reform efforts quite similar to the ones enacted around Mangarbani. The Forest Department, however, argued that this was actually forest land. There was no question of compensation; the petitioner simply requested that the government not "forcefully dispossess" his family.

Eviction and Construction

The court is waiting to hear from the Revenue Department, which is in charge of land records, before making its final decision.[92]

Rich Publics, Poor Publics, and a Fractured State

As this case suggests, the 1994 and 1996 notifications did not just affect village landowners. In Gujjar-dominated villages (which cover most of the Southern Ridge), other castes and communities had long struggled to improve their position, aided by the occasional government scheme, before being blindsided by the aforementioned greening initiative. Further, as the city sprawled outwards, formerly rural areas of Delhi drew their share of migrants, both rich and poor. The differential treatment of these migrants, especially after M. C. Mehta's prodding and the resulting 1996 notification, makes it difficult to believe that the state is always working for an unbiased, universal "public interest". However, the Ridge's post-1996 history also suggests that the state is deeply fractured, and its fractures open up a range of possibilities for the privileged and the oppressed alike.

After the notification, elite settlements on the Ridge generally remained untouched. There are two iconic examples of this. One is the Polo Ground run by the President's bodyguards on the Central Ridge. After a public outcry, this was briefly shut down, but it re-opened once the media's gaze had shifted. The other is Sainik Farms, Delhi's most famous rich "illegal" neighborhood. As the name implies, the neighborhood is built on land that was zoned in the Master Plan as agricultural, but its super-rich residents are hardly farmers. Portions of Sainik Farms are on the Southern Ridge, but its residents have little reason to worry; government officials know not to touch this well-connected locale.

On the other hand, working class settlements on the Ridge were targeted after the 1996 notification, and most, including two of the three settlements surrounding the Bhatti mines, were demolished, although it sometimes took many years for the bulldozers to actually arrive. But as the previous chapter's exploration of Bhagirath Nagar shows, working-class settlements were not totally defenseless. In Bhagirath Nagar, a local politician, connected to a national political party, opposed a Supreme Court ruling, and its implementation by municipal officials.

Of course, a divided state is hardly a new phenomenon. Since the early days of the Sultanate, state power in Delhi has been riven by competing factions. But in contemporary politics, as state power increases, so too do the instances of fragmentation. This is especially pronounced in a place like Delhi, the center of national politics and bureaucracy. The tensions are myriad; local officials resent the influence of bigger players, whether imperial (in British times) or national (in postcolonial times). This applies equally to Beadon and to the current Chief Minister of Delhi, Arvind Kejriwal. Different departments vie for influence. The judiciary scolds the executive branch, and legislators defy the wishes of both, sometimes under the influence of popular pressure, sometimes under the influence of bribes (and often under the influence of both).

The administrative structure of the Ridge ensures that it will fall prey to these rivalries and tensions. Although the entire Ridge is now nominally under the control of the Ridge Management Board, its everyday administration is still divided between several governmental organizations, namely: the DDA, the L&DO, the Forest Department, the Revenue Department, the Army (never far from the Ridge), the Municipal Corporation of Delhi, the New Delhi Municipal Council, the Ministry of Home Affairs (MHA), the Railways, the Central Public Works Department, and the Sports Authority of India (which controls the Asian Games-era shooting range). Many of these only control small patches of the Ridge (the MHA, for instance, owns a scant six hectares); the biggest owners are the Forest Department and the DDA, two agencies with very different visions for the Ridge.

One sign of the state's fractured nature is that, over two decades after the sweeping Ridge notifications of the mid-1990s, the settlement process is still not completed; the Ridge's boundaries have not been definitively demarcated and residents' claims to compensation have not been fully processed. Daunted by the enormity and complexity of the task, and often aware of the financial incentives (in the form of bribes and kickbacks) of delaying the process, various government officials and departments have found reasons to slow down the process and shift to blame to other organizations. The judiciary has repeatedly intervened in the matter, asking the government to comply with its orders and finish the settlement process so that the Ridge can finally, after more than a century, become a fully Reserved Forest.

The National Green Tribunal (NGT), a special governmental body set up in 2010 to expedite environmental cases, has added its voice to the cacophony, chastising the government for its inaction and demanding that the settlement be completed. The Forest Department, never very powerful in the Delhi hierarchy, has been energized by the NGT's prodding; in 2012, it set up a Special Task Force for surveying the Ridge and demarcating its boundaries. Forest Department officials have been especially excited about using new technology to carry out the survey; this includes using digital maps from the Survey of India, along with Geographic Information System (GIS) software to draw precise boundaries.

These digital tools have been complemented with real-life, ground-level surveys, especially after the NGT's further intervention in 2013, decrying the "unmanageable situation" on the Ridge. The Forest Department was already well-aware of poor "encroachments" like Sanjay Colony. However, during their surveys, they "discovered" richer settlements on the Ridge as well, which had proliferated with the tacit support of earlier generations of officials. In April 2014, Forest Department officials knocked down the boundary wall of a luxurious "farmhouse", claiming that this was Reserved Forest territory.[93]

The owner of the farmhouse was shocked at the audacity of the Forest Department. He, along with several other farmhouse owners, petitioned the Lieutenant Governor, asking him to reverse the NGT's orders. The Lieutenant Governor did not act immediately, but in September 2014, the Delhi chief secretary ordered the Forest Department to turn over its maps and surveys to the Revenue Department, so that it could cross-check the new demarcations with its old land records.[94] As of 2019, this process has not been completed; the settlement saga continues.[95]

This has been a reprieve for the rich and the poor alike, although, like most state actions in Delhi, it will likely benefit the former more than the latter in the long run. To wit: after the farmhouse owners complained, but before the Ridge case was officially transferred to the Revenue Department, the government demolished a working-class settlement in the Ridge village of Aya Nagar, bulldozing an estimated 250 buildings.[96]

Forest Department officials, despite their frustration with having the Ridge case transferred away from them, are still confident about their surveillance and mapping methods, echoing the British faith in sharp boundaries and strict demarcations of space. This language is reinforced by the Ridge Management Board, including its civil society members. Sunita Narain, a Ridge Management Board member and head of the Centre for Science and Environment, noted that the board "will do annual surveys, using remote sensing and spatial technologies, to assess tree cover under each agency. Any evidence of encroachment or degradation from the baseline will be severely penalized."[97] However, without addressing the underlying power dynamics that shape land use in Delhi, these increasingly invasive tools will simply be one more part of an arsenal of injustice, applying a supposedly neutral law unequally and targeting those with the least power.

The ongoing contested demarcation of the Ridge shows both the power of the state and its peril. Laws like the Indian Forest Act and the Delhi Land Reform Act, and bodies like the National Green Tribunal and the Supreme Court, have tremendous power in creating categories that then have life-and-death consequences for the bulk of the population; these agencies define what "public interest" will entail (even if it involves mass evictions and dispossession), what land is a "forest" (even if it has no trees) and what qualifies as "surplus" (even if it has myriad uses for local populations). But these categories are not as airtight as their proponents would like. The fissures in the state also break the spell of these categories.

Precarious Lives on the Ridge

For many working-class residents of Delhi, especially those living on the Ridge, the tensions within the state create conditions for a troubling, cyclical existence. A brief snapshot of one working-class Ridge-top neighborhood suggests the trials its residents are forced to face.

Not far from Qutb Minar, just behind Mehrauli's famous *jharna*, there is a sprawling informal settlement. I used to walk by this neighborhood, on the way from my flat in Mehrauli to the nearby Chhatarpur Metro Station. Along the main road were a lively set of stalls. I would often stop for a glass of chai or roohafza, depending on the weather, or, if I had more time, for kebabs or pakoras or chole bhature. With its bright lights and its buzzing crowds, this market and its adjoining settlement made it feel safe to return home even late at night, unlike the more lonely route to Mehrauli via Qutb Minar Metro station.

But for many government officials, the settlement was seen as an encroachment. It was first demolished in 1972, as a sort of prelude to the massive Emergency demolitions of the mid-1970s. However, with tacit support from other government officials, who saw the settlement as a source of votes, of power, or of bribes, the neighborhood was rebuilt, until the next demolition drive in 1995, following the major Ridge notification the previous year. Again, the neighborhood regrew, only to face bulldozers once more in 2012, when the National Green Tribunal began bearing down on other government agencies. Finally, in December 2014, several months after the Forest Department's demarcation drive had supposedly been suspended, the neighborhood was demolished yet again.

A friend and I stopped at a chai shop a few days after the most recent demolition to ask about what had happened. A chaiwala sat with several customers. They were surprisingly nonchalant, even though the chaiwala and one of the customers had lost their homes in the destruction. They described the events with a deadpan gallows humor: how the government had sent notices about the demolition, but no one believed them; how they didn't have time to take their belongings before the bulldozers leveled their houses; how people were contemplating going back to their ancestral villages, but in the end, never would. We asked if people would reconstruct their houses. "No," a man said. "It is finished."

But he was wrong. Phoenix-like, the neighborhood has risen again. Houses have been rebuilt, shops re-established along the busy main road. Life goes on, though precariously. In the shadow of the Qutb Minar, the state still hovers, permeable, contestable, but all too real.

4

SURPLUS
Production, Consumption, Speculation

New Rich, Old Rich (New Ridge, Old Ridge)

Mehrauli's Ridge-top *jharna*, and the troubled settlement beside it, lie south of Qutb Minar. Directly north of Qutb Minar is a once-luxurious, now-fading building called the Qutb Colonnade. Built in British colonial style, the colonnade was the center of an attempt to gentrify Mehrauli in the 1990s. A key player in this effort was Bina Ramani, the socialite and fashion designer who had earlier engineered the conversion of nearby Hauz Khas Village from a sleepy set of buildings and ruins to Delhi's hippest party hub.

Ramani ran a restaurant called "The Tamarind Court" in Qutb Colonnade. On an April night in 1999, Ramani threw a lavish party for her husband. She claimed it was a private party, but anyone with enough money could buy entry to the event. Ramani had set up an unlicensed bar in the restaurant and had hired models as bartenders.

The party continued long after the restaurant's closing time. After midnight, Manu Sharma, the son of an influential politician and the heir to a lucrative sugar mill business, strode into the restaurant, approached the bar and asked for a drink. The bartender he confronted, a 34-year-old model named Jessica Lal, told him that they had just run out of alcohol. Sharma took this as an affront and again demanded a drink. When Lal again refused, Sharma took out a pistol and fired it into the air. His pride bruised by the refusal and his temper flaring, he asked for a drink a final time. Lal said no. Sharma shot her in the head, killing her instantly.

This tragic killing, with its senseless violence and its high-profile players, was widely covered in the media. It was even made into a Bollywood movie, *No One Killed Jessica,* which focused on the aftermath of the shootings, when most of the famous witnesses turned hostile and refused to testify against Sharma and his friends. Many alleged that Sharma's powerful father had paid huge sums to convince the witnesses to change their testimony. In 2006, all of the defendants in the case were found not guilty. This led to such an intense public outcry that the case was retried. This time, Sharma was found guilty of murder and sentenced to life imprisonment.

The Jessica Lal case captured the public imagination because it revealed the seamy underbelly of Delhi's post-liberalization era and its increasingly crass celebration of wealth, status and celebrity, a celebration that often found its most conspicuous forms in the luxurious buildings of the Southern Ridge.

But arguably, Delhi's tryst with material wealth can be traced much, much further back, not to liberalization in 1991, but to the flowering of Paleolithic production. As discussed in Chapter 1, an immense profusion of stone tools were manufactured on the Ridge during Paleolithic times. Some archaeologists have theorized that such tools were the first commodities, since they could be produced in large numbers in a relatively uniform way, and thus hoarded and traded.

This remains in the realm of speculation. Even if stone tools were part of an early, basic trading system, the livelihood strategies of hunter-gatherer tribes did not allow for accumulation on a scale that would enable some people to live off the surplus produced by others. This would only come with the arrival of agriculture and (more importantly for the Ridge) pastoralism.

The rise of surplus also led to the rise of the state; this chapter and the previous one are thus intimately interlinked. For much of Delhi's history, those who were extracting surplus and those who were building states were one and the same. In the case of Delhi's first Ridge-top fortresses, the surplus-extractors and state-builders were Rajputs, who had emerged from the relative isolation of tribal life and had become increasingly powerful, forming alliances, levying taxes and building kingdoms.

Adventurers and plunderers from other parts of the Arid Zone were drawn to India precisely because of its well-developed states and their carefully managed surpluses. The Sultans, taking control after Ghori's incursions, established Delhi as a prominent center of wealth, which was then targeted by new rounds of conquerors from the north, from the Mongols to the Mughals. For the Sultans setting up a base at the edge of the Arid Zone, this was, as we've seen, largely a process of balancing and controlling the agricultural and pastoral populations in the surrounding regions, utilizing the surpluses they produced while putting down rebellions from both within and without. The Mughals, who consciously positioned themselves as heirs to the Sultanate legacy, performed a similar balancing act.

The British came to India as one more power drawn by India's fabled wealth. The British, unlike earlier rulers, initially came to the subcontinent as traders, in the guise of the East India Company. Here too, though, the state was far from absent; the East India Company depended crucially on the monopoly status granted to it by the British crown. The lines between state and surplus became further blurred when the Company started to behave less like a trading firm and more like an empire. As its territorial ambitions grew, Delhi, that old seat of power and wealth, was inexorably drawn into its range.

Since then, state building and surplus production have co-existed in an uneasy, complex alliance, often formally separated, but always intersecting. The nature of the alliance shifted, as the mercantile capitalism of early Company rule transitioned to the industrial, imperialist capitalism of the late British period, which in turn gave way to the socialist-inflected developmentalism of the early post-Independence period and finally the re-embrace of a naked capitalism in contemporary times. As this chapter seeks to argue, the Ridge offers a useful microcosm for analyzing the shifting nature of this alliance.

If the state has crucially depended on the Ridge, so too have the accumulators of surplus. The Ridge has provided much-needed mineral resources (for stone tools, for fortress walls, for concrete, for road foundations) as well as much-needed ecological ones, including the grasses and shrubs that supported pastoralism for centuries. Since the colonial period, when the British introduced the radical idea that land is simply a commodity like any other, the Ridge has provided a more nebulous source of worth; increasingly, it is prized not for its geology or ecology, but simply for its value in a lucrative real estate market.

The Ridge is not just a site for the production of economic surpluses. It has also emerged as a favored locale for the (conspicuous) consumption of surplus, as the tragic story of Jessica Lal indicates. This function of the Ridge is not disconnected from its ecology. With its tough, stony soil, the Ridge was never a favored agricultural site and thus was less densely populated than fertile zones in Delhi. The Ridge's relative isolation proved appealing to elite revelers looking to distance themselves from the masses.

This trend has quite a long history. As we have seen earlier, in the Sultanate period, quartzite lodges sprang up on isolated parts of the Ridge and were used for the quintessentially elite pursuit of hunting. The most prominent of these is Firoz Shah Tughlaq's lodge on the Northern Ridge, built as a regal retreat where he could recover from the emotional blow of his son's death. There are several on the Central Ridge as well, which are now part of the picturesque scenery of Lutyen's Delhi. On the Southern Ridge, in Mehrauli, the Mughals built elaborate recreational monuments like the *jharna*, where they enjoyed their monsoon holidays and escaped from the dense crowds of Shahjahanabad.

With its dual roles, supporting both the production and consumption of surplus, the Ridge has been a key part of Delhi's many economic transformations. These transformations bring our attention to much larger networks of trade and governance, as the Ridge has once again been a key node in regional, national and global circuits and flows. These have not been merely economic or political phenomena. The changing fortunes of the Ridge have unleashed strong, often violent passions: a volatile mix of desire, pride, jealousy and rage. An economic history of the Ridge is thus also the history of rivalries, power struggles, conspiracies and murders.

Making Delhi Safe for (British) Business

We already have some indication of this turbulent history. In the last chapter, we met William Fraser, who was assassinated while riding his horse along the crest of the Northern Ridge. His untimely demise came at the hands of a Mughal nobleman who resented the interference of a new political power and who feared the loss of his aristocratic wealth. But before Fraser succumbed to an assassin's bullets, he had lived a full, violent, eccentric, profitable life, in which his Ridge-top mansion was a crowning achievement. Fraser serves as an apt symbol for the rough-and-ready mix of state-building and surplus accumulation on the Ridge in colonial times. A government official with a wild streak, he helped clear the way for British power in Delhi, while amassing his own riches and displaying them on the heights of Delhi's hills.

Fraser rose to prominence in the last days of the East India Company's expansionary phase, in which mercenaries and military officers built up the company's holdings across wide swaths of the subcontinent, in areas widely seen as hostile and lawless. Fraser was sent to Delhi to be the Resident's Assistant in 1805, just two years after the British had taken control of the city. The Resident's main responsibility was to control the Mughal empire by proxy, but he and his assistants also had to deal with a Delhi region that was still in a state of tumult. Fraser's first task was to tame Delhi's turbulent hinterlands, just as Balban had done centuries earlier when he had seized control of the disintegrating Sultanate.

Fraser's adventures in Delhi, military and otherwise, have been chronicled by William Dalrymple, whose wife is Fraser's descendant. Dalrymple was lucky enough to chance upon a trunk full of letters, documents and paintings in the Fraser ancestral home in Scotland while he and his wife were vacationing there. Dalrymple reproduces a letter from Fraser's brother Alex, which draws parallels between the rugged Scottish highlanders and the nomadic Gujjars tribes of the Delhi area, both known for their pastoral pursuits. The letter describes William as being "surrounded by Goojurs, formerly Barbarian, now like highlanders; independent to equals, fiery and impetuous, but faithful and obedient".[1]

After several years of military successes, Fraser began the project of building his luxurious Ridge-top mansion. He spent many years and poured much money into making it. He chose the location carefully: it was built on the very spot where Timur had camped centuries earlier, as he prepared for his raid of Delhi.

While Dalrymple does not choose to ignore Fraser's violence and mercurial nature, he prefers to dwell on Fraser's embrace of courtly life, his mastery of Persian and Urdu, and the lasting bonds he developed with Mughal nobleman. For Dalrymple, Fraser and other "white Mughals" represented the "hopes of a happy fusion of British and Indian culture".[2]

It may be questionable to romanticize those who won loyalty through a reign of terror. But Dalrymple is certainly right in identifying Fraser, and other British officials like him, as a dying breed in India. They represented an appreciation for Indian languages, customs and cultures (at least in their elite manifestations) and a warm respect for their Indian rivals and allies, which was rare by the time Fraser died and was definitively swept away by the bloody tide of 1857.

Dalrymple attributes this shift to an increasingly racist attitude on the part of the British, which was complemented by an increasingly rigid adherence to Christian doctrine. But this is only a partial explanation. The changing attitude towards India and Indians was also a product of massive changes in the global economy, and Britain's leading role in these changes. When the East India Company first ventured into India, it was one among many European proponents of mercantile capitalism. By the time of the 1857 Uprising, the age of industrial capitalism was well underway, and the United Kingdom was by far its most powerful proponent.

As industry, and in particular, the textile industry, expanded at a furious rate, with the United Kingdom at its center, countries like India suddenly found themselves at the periphery. With improved technology and machinery, Britain could produce massive amounts of textiles at extremely low prices. Further, belying their "free trade" rhetoric, the British disproportionately taxed non-British textiles and engaged in the physical destruction of handlooms in their empire, hence undercutting competition from India, which was long known for its hand-crafted textiles. At a political level, the Mughals, and other kingdoms of India, were no longer seen as rival powers, to be fought on the battlefield and viewed on more or less equal terms. They were seen as subordinate members of an increasingly integrated global economy.

The two factors suggested by Dalrymple—increasing racism and evangelical Christian fervor—should also be seen in light of these tectonic shifts in world affairs. While scientific and religious developments were not simply reactions to economic change, there is no doubt that these three realms influenced each other to a great degree. Thus, the newly subordinate economic position of countries like India was rationalized through the increasingly influential "science" of race, from phrenology to eugenics. At the same time, social churning and widespread instability in the wake of the Industrial Revolution found many outlets, including religious revivalism.

By the time of Fraser's death on the Ridge, these interlinked developments had brought an end to the age of the "White Mughals". But Fraser should not simply be seen as a man increasingly out of place in his era. It was precisely through the military efforts of Fraser, and others like him, that the East India Company established the conditions for a new era of colonialism. Fraser's efforts to tame Delhi's hilly hinterlands, and their reliably rebellious nomadic communities, laid the groundwork for the later settlement efforts by J. R. Maconachie and Henry Beadon, which sought to instill the values of capitalist agriculture in the Delhi region.

As we saw in Chapter 2, these attempts at "settlement" had wide-ranging impacts on the Ridge, where village commons were threatened, pastoral pursuits became increasingly precarious, and plots of land were increasingly seen as commodities. And while these trends reached their peak in the post-Uprising period, they had already gained momentum during Fraser's tenure as Resident.

Mughals, Metcalfe and Murder on the Southern Ridge

The man who became resident after Fraser's death, Thomas Metcalfe, took advantage of these new economic conditions, buying up huge plots of land in Delhi, while also embodying the new British conservatism that Dalrymple so disdains. Instead of embracing Mughal dress, as some of his predecessors had, "he arranged that his London tailors... should regularly send out to Delhi a chest of sober but fashionable English clothes."[3]

However, Metcalfe appreciated the symbolism and the splendor of Mughal tradition. In this, he foreshadowed the builders of New Delhi: he wanted to appropriate the symbols of Mughal power, not as an equal (like Fraser had tried to do), but as a superior successor.

In doing so, he turned to the Southern Ridge, and more specifically to Mehrauli, that favored haunt of Mughal pleasure-seekers. He combined courtly Mughal traditions with his own ideas of luxury, drawn from traditions of British landscaping and aesthetics. He, like other British officials, imagined the Ridge as a miniature hill station, a welcome relief from Shahjahanabad's urban congestion. In addition to his palatial house at the foot of the Northern Ridge (which rebelling Gujjars burned down in 1857), Metcalfe bought a tomb in Mehrauli and converted it into a monsoon getaway.

The context of this purchase suggests that economic unrest was already rippling across the Delhi countryside. As Metcalfe's daughter Emily reports, "The family to whom it belonged had become impoverished and had handed over this tomb, as their only available asset, to the banker to whom they owed a large sum of money. He wished to sell it, and so my father bought it."[4]

By Mehrauli terms, the tomb is not very old. It was built around 1610 to house the grave of Quli Khan, a foster brother of the Mughal emperor Akbar. Metcalfe occupied the tomb and renamed it Dilkhush, or "The Heart's Delight". He added pavilions and terraces, as well as several new rooms. In its finished form, it included a master bedroom, a smaller bedroom for Emily, a library, a drawing room, a dining room, a dressing room, and several guest bedrooms. He turned another, smaller tomb nearby into a boathouse, which bordered a small pond and waterworks he had created to amuse his guests. He also rented out the house to honeymooning couples.

Metcalfe was prescient in choosing the locale of his elite getaway. Though in one sense he was following in the footsteps of the Mughals, in another sense, he was a trailblazer in his gaudy re-appropriation of the Southern Ridge's long history, a foreshadowing of the farmhouse trend to come. Unlike the Mughals, but like his postcolonial pleasure-seeking successors, he chose a very secluded spot, far away from the bustle and fervor of Qutb Sahib's shrine. His exclusive ponds and boats presaged the fenced-off farmhouses and elite swimming pools of the present day. And not least, with his side business of honeymoon rentals, he was a precursor to the booming real estate and marriage industries on the Southern Ridge.

These comparisons should not, perhaps, be taken too far. Metcalfe was distinctly British, and distinctly colonial. One sure sign of this is the series of "follies" he built around his Mehrauli compound. In architectural parlance, a "folly" is a structure built purely for decoration, with no functional purpose. Follies became popular in the eighteenth century in British and French gardens, and often incorporated classical Greek or Roman elements. Others were more whimsical or exotic, suggesting a fascination with the non-Western world that was opening up through trade and colonial expansion. Egyptian pyramids and Chinese temples were thus recreated on the lawns of English nobility.

Bringing this idea to India, Metcalfe built a series of follies that he and his guests could see while relaxing, boating and dining at Dilkhush. Although Dilkhush was already surrounded by tombs and monuments in various states of decay, Metcalfe wanted to enhance this melancholy effect through strategically placed follies. Some were relatively nondescript, such as a plain quartzite tower build atop a pile of rocks in the distance. Some were more elaborate, such as a pavilion built in Mughal style. And some were just bizarre, including pyramids and ziggurats.

Metcalfe seemed to pretend that these follies were just part of the landscape, writing of Dilkhush,

> The ruins of grandeur that extend for miles on every side fill it with serious reflection. The palaces crumbling into dust... the myriads of vast mausoleums, every one of which was intended to convey to futurity the deathless fame of its cold inhabitant, and all of which are now passed by, unknown and unnoticed.[5]

This evocation of the subcontinent's storied past had a clear instrumental role: it was used to justify colonial intervention, and to claim that the British had stumbled upon a degenerated empire that needed resurrection. While Metcalfe and his ilk meditated upon the transitory nature of power and empire, they had no doubt that, in the present moment, their empire was in the ascendant. Their recognition that glory inevitably fades did little to stop them from constructing monuments to their own glory, in both small ways

(Metcalfe's follies) and large (the building of New   everal decades later). For Metcalfe, the Ridge around Qutb   with its rocky vistas and its high density of ruins, was the   location for his peaceful, melancholy retreat, and he was happy to share this experience with other British visitors, for a price.

But Metcalfe, like Fraser before him, was unable to escape the tumult that British rule had brought to Delhi. His demise, too, was linked to Mughal inheritance disputes. But while Fraser fell afoul of a minor nobleman, Metcalfe attracted ire on a grander scale: his enemy was the Mughal Emperor, Bahadur Shah Zafar, and his favorite wife, Zeenat Mahal.

At the heart of the dispute was the question of Zafar's rightful heir. Zafar, poet, mystic, and soon-to-be leader of the Uprising, was not fond of his eldest son. Under the increasingly strong influence of Zeenat Mahal, he became convinced that his proper heir should be one of her sons. Metcalfe objected, citing the old British principle of primogeniture.

Zafar became enraged when Metcalfe, along with two other British officials, initiated talks with his eldest son, Mirza Fakhru. In 1852, Mirza Fakhru signed a secret pact with the British, which stipulated that he would be the next emperor, as long as he moved the seat of Mughal power from Shahjahanabad to Mehrauli, and as long as he formally dropped the Mughal's claim of superior status.

In 1853, Metcalfe fell ill. When the illness didn't lift, he started to suspect that he was being poisoned. This suspicion was strengthened when he learned that the two other officials involved in the secret pact were experiencing similar symptoms: weakness, nausea, vomiting. Unable to keep down any food, Metcalfe started wasting away. After several months of sickness, he died in his north Delhi mansion.[6]

As it turned out, the controversy over Zafar's heir was a moot one. The Mughal Empire ended with Zafar and with the post-Uprising retributions of the British. In 1858, the British Crown finally took control of India, officially superseding the East India Company. Delhi, for a time, became a shell of its former self. But with the growth of British New Delhi, fully backed by the imperial government, the city's prestige was soon resurrected. The Ridge too regained its role as a site for displaying power and wealth, a function that has continued to the present day.

Postcolonial Ridge Real Estate

As we saw in the previous chapter, land speculation, and the government's attempts to curtail it, were a key element in the establishment of British New Delhi. From 1911 onwards, real estate has played an increasingly central role in the production and consumption of wealth on the Ridge, leading to luxuries and architectural excesses that would put Metcalfe to shame.

To understand the rise of the real estate business in Delhi and on the Ridge, it is necessary to dig into the history of Delhi Land and Finance (DLF), the country's largest real estate developer. We have already encountered DLF; the company featured as one of the many roadblocks Sajjan Singh faced in his quest to receive compensation for the acquisition of his family's ancestral land on the Central Ridge. But this was just a minor episode in the long, storied, some would say notorious, trajectory of DLF. The history of this private company is marked by a complex relationship with state institutions, including the powerful DDA. If the DDA has been the guiding spirit of public-sector housing development in post-Independence Delhi, DLF has played that role for the private sector.

DLF was founded just before Independence, in 1946. Its founder, Chaudhary Raghvendra Singh, had studied at the elite, British-run St. Stephen's College and then joined the Punjab Civil Service in 1935. With the outbreak of World War II, he joined the Indian Army, which was still run by the British. For his prominent role in recruiting Indian officers to join the service, he was eventually named an honorary Member of the British Empire.

With the end of the war, Singh sought to reinvent himself as a real-estate tycoon. Soon after, Partition brought a flood of refugees who needed housing. Many of the refugees were from well-off urban backgrounds, and Singh's DLF was the prime supplier of new housing colonies for this well-heeled group. The company was extremely prescient, buying up large swaths of agricultural land south of Lutyens'

Delhi in anticipation, not just of the swelling refugee population, but more long-term urban expansion.

Singh and his colleagues could clearly see that Delhi's center of gravity was shifting. Before Raisina Hill became the seat of government power, the cities in Delhi had been slowly moving northward; the Southern Ridge cities of the early Sultans were replaced by Siri to its north, and then Firozabad north of that, and then Shahjahanabad north of that. Though the British briefly contemplated a site north of Shahjahanabad, their decision to locate their new city on the Central Ridge reversed centuries of northward drift.

As suggested in the previous chapter, the current city of Delhi is, in many ways, the direct descendant of the capital established by the British in 1911. Not only has it inherited its grand halls of governance on Raisina Hill, it has also kept many British laws and attitudes, and has given immense power to the DDA, the successor to the British DIT. And while the city has expanded in all directions, the city's elite have largely followed the British lead and settled further and further south.

For DLF, the land south of New Delhi was an obvious choice. To New Delhi's north was the crowded old city of Shahjahanabad, to its east was the Yamuna River, and to its west was the already-packed city extension. The south, meanwhile, was a more open landscape, dotted by agricultural and pastoral villages. DLF planned developments all over Delhi, but the south was clearly its favored location. Even today, the colonies DLF founded in South Delhi, including South Extension, Greater Kailash, Kailash Colony and Hauz Khas, are among the capital's most coveted.

DLF was also looking farther afield, to the Southern Ridge villages which were even further south than the colonies the company was rapidly constructing. These were mostly Gujjar-dominated villages eking out a pastoral existence on rocky land, sometimes as far as 30 kilometers away from the center of Delhi. In the 1950s, DLF agents approached many Gujjar landowners and bought up large tracts of their land. DLF's leaders intuited, quite correctly, that this distant, forbidding territory would eventually come into the urban orbit and thus into the world of ever-increasing real estate prices, to be valued by the city elite precisely because of its sparse population.

The decade after Independence and before the formation of the DDA was the heyday of private real estate development in Delhi, with DLF as the biggest actor, both in the city's center and at its periphery. Throughout this decade, there were widespread accusations of corruption and private-sector collusion with local government officials. Nehru viewed DLF with great suspicion, and he pushed for a body like the DDA largely to curb the power of DLF and their covert allies in the Delhi bureaucracy.

Nehru was galled by the brazenness of top Delhi officials, including the Chairman of the DIT, who bought huge amounts of land for as little as four annas per square yard, and then proceeded to sell it for eight rupees per square yard. An exasperated Nehru wrote to the Minister of Works, Housing and Supply saying, "This does seem to me rather extravagant profit.... Of all persons, surely the Chairman of the Delhi Improvement Trust should not make money in this way."[7]

In the end, the establishment of the DDA could not stop the rampant land speculation, nor the collusion of high-level government officials in shady real estate deals. It did, however, stop the momentum of DLF, as all private real estate developers were prohibited from plying their trade on DDA land. The Master Plan's strict guidelines for agricultural land also slowed DLF's growth on Delhi's fringes. For a time, Nehru's bid to limit the reach of private real estate companies like DLF seemed successful.

From the Ridge to Gurgaon

As the decades have passed, Nehru's success has appeared increasingly ephemeral. DLF has since recovered its status as Delhi's premier real estate developer. Its farmhouse complex in the Southern Ridge, DLF Farms, is now an icon of residential luxury; another DLF project, the high-end Emporio mall, has become the city's symbol of consumer luxury. The mall sits on the southwestern corner of the Delhi Ridge, in a neighborhood called Vasant Kunj. It is part of a larger commercial complex that also houses the corporate headquarters of Maruti Suzuki, the country's biggest car company and a prime symbol of the age of economic liberalization.

It is perhaps no coincidence that DLF and Maruti Suzuki sit together on the Vasant Kunj Ridge. They represent key components of the Delhi region's recent economic history: on the production side, the lucrative expansion of real-estate development and large-scale manufacturing; on the consumption side, the promotion of a lifestyle increasingly based on malls and cars. But these two companies, and the two industries they represent, are not in perfect synergy. Though they sometimes find common ground (as they do, quite literally, on the Vasant Kunj Ridge), they often find themselves in conflict, representing different visions of economic growth and urban expansion.

If the convergence of these companies is easily seen on the Ridge, the divergences are best understood by taking a brief detour and leaving Delhi altogether. The Vasant Kunj commercial complex is a relatively recent development; we will return to this site and its controversial history later in this chapter. Well before their forays into the Vasant Kunj Ridge, both DLF and Maruti Suzuki found success, and courted conflict and controversy, in less rigorously regulated climes outside of Delhi.

Specifically, both companies were drawn to Gurgaon, the village-turned-suburb whose birdwatchers rallied to defend Mangarbani in Chapter 2. Gurgaon sits just south of Delhi, in the state of Haryana. The Southern Ridge forms the boundary between the city and its new suburb. Companies like DLF had to temporarily forsake the Southern Ridge and forge further ahead, before they could loop back and make their triumphant return.

For both Maruti Suzuki and DLF, 1981 was the key year for initializing their Gurgaon growth. But for both companies, this development had a long prehistory. Tracing the twin storylines of these companies reveals much about the changing nature of the state/market relationship in India, the country's increasing entanglement with global markets, and the wide-ranging consequences of these changes for the human and non-human inhabitants of Delhi and its Ridge.

Gurgaon Detour, Part I: Maruti

Maruti started as a pet project of Sanjay Gandhi, Jagmohan's ally in Emergency-era slum demolitions. After completing an internship with Rolls Royce in England and returning to India in 1968, Sanjay convinced his mother Indira Gandhi, then early in her career as Prime Minister, that the government should grant a license to a private-sector company for the production of a small, affordable car. This was the height of the so-called "license Raj", with the government attempting to control the economy by maintaining strict regulations on what company could produce which products.

Indira Gandhi's nepotism was already well-established by this time, and no one was surprised when a company founded by Sanjay himself, called Maruti, received a license to produce India's first indigenously made low-cost car. Soon after, the Chief Minister of Haryana, Bansi Lal, gave Sanjay a personal tour of Haryana, and in violation of government land policies, let Sanjay pick the plot of land he wanted for his upcoming car factory. Bansi Lal sold him about 300 acres of land in Gurgaon at an extraordinarily low price.[8] But even with all these favors, Sanjay was not able to produce a working prototype for the car. After Emergency, when the Janata Party came to power, they scrapped the company, as one of many moves to erase Sanjay's corrosive legacy.

Many point to this debacle, and other cases like it, as a sign of the failure of India's post-Independence economic model, some-times referred to as "Nehruvian socialism". However, as the sociologist Vivek Chibber has shown in his account of post-Independence industrial policy, there was little that was truly socialist about the Nehruvian economy. Capitalists in India had no problem with a planned economy as long as that meant state support for industry, but no state tools for disciplining capital. The capitalist lobby suc-ceeded in crippling the main elements Nehru proposed to reign in big business (such as the Planning Commission and the Industrial Disputes Resolution Act), thus bending the planned economy to their interests.[9] Generational change also exacerbated the weaknesses of this economic model; whereas Nehru truly believed in the importance of a planned economy, Indira Gandhi exploited it more cynically. She used it not just to meet her son's whims, but also to shore up support as the dominance of Congress faded.

However, Indira could also see that the world was changing. Especially after the global economic crisis of 1973, governments around the world were abandoning their earlier faith in capitalist state planning, and were emphasizing the need for looser regulations, increased privatization of industry, a denser network of global trade, a reduction of import tariffs, and a host of related measures.

Although India would wait until 1991 to fully embrace the neoliberal policy package, it started to move tentatively in that direction during Indira Gandhi's later years. The rebirth of Maruti in the 1980s was one sign of this. Whereas Sanjay's project emphasized indigenous production, the new Maruti Udyog, though a public-sector company, partnered with the Japanese firm Suzuki. The new company, founded in 1981, began production in Gurgaon two years later. Unlike Sanjay's company, this iteration of Maruti quickly developed a low-cost car and marketed it with great success. The company has since built on its early triumphs, and it still dominates the car market in India, even in the face of increasing competition. At the same time, the company has become increasingly privatized. Initially, Suzuki owned a minority stake in the company; as the economic winds shifted, this increased to 50 percent. By 2007, the government had completely disinvested in the company.

Maruti Suzuki has been trumpeted in the business press as a success story for the new Indian economy. The growth of the company has been one of the engines driving Gurgaon's expansion. The company also had a broader symbolic impact, as its low-cost car was one of the most visible signs of globalization's prehistory in India and a shift away from an earlier ethos. The Nehruvian era had focused, at least in rhetoric, on major industrial projects that would benefit the nation: dams, steel factories, coal mines, power plants. Along with this was a focus on saving and austerity. The good citizen would not spend profligately on consumer goods but would save up money for the good of the country.

The Maruti Suzuki 800, the company's first vehicle, was a harbinger of change. Although the true watershed only came in 1991, Maruti Suzuki was an early sign that new values and trends were creeping in: consumerism, personal loans, international branding, private vehicles. Good citizens now bought the Maruti Suzuki 800, which was portrayed as a "people's car".

Beneath this branding exercise lurked traces of liberalization's dark side. True, the Maruti Suzuki 800 was much less expensive than other cars manufactured in India, such as the famous Ambassador, long considered a powerful status symbol. It was thus embraced by the growing middle class, who enjoyed a newfound mobility (both metaphorical and literal). But it was still out of reach for the vast majority of the population, who could not afford the "people's car" and found themselves rhetorically removed from the category of "people".

Such language betrays an increasing indifference to those left behind by the new economy. The post-liberalization age has witnessed sharp increases in inequality. A select few have seen their fortunes soar, and a much larger group now has the opportunity to buy an ever-growing range of imported goods, but those at the bottom of the pyramid face an increasingly precarious life. Worse, their plight finds little space in the imagination of a media and a middle class that imagine themselves as representative of "the people".

This shrinking of the social imagination has an ecological impact as well, as people want to consume more but often have little concern about the environment, except for their immediate surroundings. It is no coincidence that the Maruti Suzuki craze emerged at the same time as the environmental PIL phenomenon; the association of "the people" with the numerically small, but symbolically important, middle class is an important part of PIL logic. This trend also underscores the hypocrisy of a Delhi-centered environmentalism that embraces the rise of the automobile while at the same time decrying industrial pollution, especially because the cars are being produced in Delhi's backyard.

If the environment suffered with the rise of Maruti, so too did the company's workers. Corporate profits have soared, but workers' wages (adjusted for inflation) have fallen steadily over the past two decades. As wages have fallen, work has also become more precarious, as permanent workers are increasingly being replaced by those on temporary contracts.

Though the state gradually (and then completely) withdrew its stake in Maruti, it remained involved in other ways. Since Independence, the state has played a key role as an arbitrator in labor disputes. Although this blunted the militancy of labor struggles, it had, for decades, given workers and unions in the formal economy a basic level of support and stability. However, with the coming of neoliberal policy reforms, the state has more and more taken the side of managers against workers.

This can clearly be seen at Maruti Suzuki, where workers long struggled to form a union that was not under the thumb of the management. They were opposed at every turn by the company, with managers intimidating and often firing the more vocal workers. In 2012, after 13 years of on-and-off agitation, a worker-controlled union was finally registered. But later that year, simmering tensions between management and workers boiled over into violence that left both workers and managers injured, and one manager dead. The mainstream media, along with the management, quickly pegged the blame for the incident on the workers. Later reports, however, revealed that the management had brought in hired goons to incite violence, and the manager who died was, mysteriously, the one most sympathetic to the workers' demands. The management subsequently enlisted state support, using the incident as an excuse to torture and jail hundreds of workers, while firing 2,300 workers.[10]

In January 2014, a group of terminated workers and their families, along with supporters from political and civil society organizations, embarked on a 15-day "Jan Jagaran Yatra" ("Journey for the Awakening of the People"). They traveled on foot from Kaithal, Haryana, where over 100 workers were still languishing in jail, to Delhi. On the penultimate day of the march, the protesters stopped outside the gleaming Maruti Suzuki headquarters on the Vasant Kunj Ridge. Before they could get close to the headquarters though, they were stopped by a line of heavily armed policemen. The police had received instructions to prohibit the group from raising slogans outside the headquarters, because it would offend the sensibilities of the corporate managers sitting inside the building. Yet again, the victims of the new economy were quite literally kept out of sight.

Gurgaon Detour, Part II: DLF

Despite its history of repressing workers, and despite its contribution to Delhi's epic pollution levels, Maruti Suzuki remains a media darling. DLF, on the other hand, has received more critical scrutiny.

This cannot just be due to DLF's notorious collusions with government officials; Maruti Suzuki has relied on state backing since its earliest days, and even in its private-sector guise, it continues to use the state to terrorize workers. Perhaps DLF has received more media criticism because, unlike Maruti Sukuzi, it has faltered in recent years, losing lawsuits and hemorrhaging money at an alarming rate. Or perhaps DLF is targeted because it reveals an unseemly, but thoroughly necessary, side of capitalist growth.

Another hypothesis: DLF finds itself in the spotlight because it has put itself there, for better or worse. Much more than Maruti Suzuki, DLF has boasted of its role in transforming Gurgaon. And it is not an empty brag: DLF's vision for Gurgaon has changed it from a sleepy farming village to a prime location for multinational corporate offices as well as a luxurious residential zone. DLF has aligned itself with a particular vision of urban growth, one driven by private real estate development, private infrastructure, international investment, business outsourcing and information technology (IT). This has become the dominant vision of Gurgaon, even as it sits in uneasy coexistence with a massive automotive industry. And it is this urban imagination that has increasingly been projected onto Delhi and the Ridge.

But, as we saw, this vision began with DLF's retreat from Delhi. This takes us back to the 1970s, when Chaudhary Raghvendra Singh's son-in-law, K. P. Singh, took charge of the company. With the Delhi real estate market officially closed, the elder Singh had become so discouraged that he instructed the younger Singh to sell his share in DLF for a mere ₹27 lakh. K. P. Singh, the dutiful son-in-law, was on the verge of selling, but, as he tells it, a conversation with DLF's financial adviser changed his mind. Instead, K. P. Singh began an aggressive campaign of buying up agricultural land in Gurgaon.

As the legend goes (and as K. P. Singh himself tells it), the new head of the troubled company was lazing in a charpoy in the scrubland of Gurgaon under the shade of a tree on a hot summer day. A Jeep screeched to a halt in front of him, and its owner jumped out of the car and asked for assistance. Luckily for Singh, the owner happened to be Rajiv Gandhi, who had officially entered the political scene after his brother Sanjay's untimely death. Singh brought some water to cool the overheating Jeep and helped get the vehicle running smoothly again. Gandhi asked Singh what he was doing in the middle of nowhere, to which Singh reportedly replied, "Dreaming of a new city."[11]

Though Gandhi was Nehru's grandson, he had not, apparently, inherited his grandfather's disdain for DLF. Not long after this chance meeting, the laws for acquiring land in Haryana changed to become much more favorable to real estate developers. In the absence of state planning and with few regulations to hold them back, DLF drove the growth of Gurgaon. The company also diversified from its role as essentially a land buyer, and has built exclusive gated communities, mega-malls, golf courses, offices and cinemas.

DLF's Gurgaon dominance began in 1981, when the company became the first to receive a license to develop property in the area. Until then, Gurgaon had been relatively ignored by regional planners and real estate developers alike. Delhi's 1962 Master Plan, which included ambitious regional planning goals, simply noted that Gurgaon is "handicapped for want of good water sources and only a modest growth is contemplated".[12] This is largely due to the configuration of the Aravallis south of Delhi. The slope of the hills channels most of the area's water into a watershed to Gurgaon's east, towards the lushly-forested Mangarbani and the now-industrialized Faridabad.

But DLF had something much more ambitious in mind than "modest growth"; it envisioned a wholesale transformation. To do this, the company first had to acquire a considerable amount of land. If the British first pushed the idea of land as a commodity in India, companies like DLF have taken this idea to its logical conclusion. However, in doing so, such companies have been forced to deal with the long, complex histories of land use in the Delhi region, including the role of the state in introducing various land reforms and regulations, and the increasing fragmentation of land as families divide up their properties. DLF had to iron out these complexities. Only then would land in Gurgaon be legible to buyers as a simple commodity that could be purchased like any other.

In this process, K. P. Singh acted as a middleman between rural landowners and urban business interests. Though the banning of private real estate development in Delhi had hurt DLF, it was still a company that planned and strategized on a grand scale. To attract both individuals and companies to Gurgaon, DLF needed to assemble large plots of land, which meant negotiating and striking deals with many different farmers. Singh stressed his rural roots when meeting farmers; he often went to them wearing a dhoti, along with a military beret to connect with the farmers' pride in their martial traditions.

But acquiring the land was only the first step in the process. The vast majority of this land was zoned as agricultural, just like the Master Plan-mandated Green Belt in Delhi. But unlike in Delhi, where a nominal adherence to the Master Plan made it necessary to develop subterfuges like "farmhouses" (which were bereft of farming), land uses could be changed more easily in Haryana. To do this, companies needed to obtain a Change in Land Use (CLU) issuance from Haryana's Town and Country Planning Department. There were many rules and regulations regarding CLUs, but these were easily bent and broken by a network of bureaucrats, politicians and real estate developers. DLF has long been dogged by accusations of its collusion with top government officials but, precisely due to its political and economic power, these charges have never stuck.[13]

Once they acquired the land and the requisite CLUs, DLF needed to find buyers. Initially, the company focused on domestic buyers. Here, DLF was helped yet again by the murkiness and moral ambiguity of the real estate world. It's a very poorly kept secret that real estate is a preferred storehouse for black money, which India generates on an epic scale. According to one study, India has over $1.5 trillion in black money, more than the rest of the world combined.[14] In the media, much is made of the black money that finds its way to banks in Switzerland and other financial havens. But many have discovered a simpler solution: investing in domestic real estate, where black money is happily accepted.

Such investments supported DLF's early Gurgaon growth. As the years went on, though, DLF's vision became increasingly global, especially following the 1991 economic reforms. In keeping with the spirit of those days, DLF turned its attention to multinational firms, the IT industry and the service sector more broadly, in an attempt to brand Gurgaon as a "world-class city". In this, there was once again a convergence between DLF's business interests and the national vision promoted by Rajiv Gandhi and his successors. Like his brother, Rajiv was enamored of technology, but he was more conscientious and methodical about its introduction in India. He continued with the technological trends started by his mother, including automobiles and televisions (which had gotten their big boost with the Asian Games). But he added an emphasis on newer technologies, including computers and networking.

As the years went on, the technological focus settled on IT outsourcing, which even today has immense symbolic importance, both domestically and internationally, despite its relatively small contribution to the Indian economy. DLF played a major role in bringing this trend to Gurgaon. Again, K. P. Singh used his personal connections and networking abilities, this time on an international level. In the early 1990s, Singh befriended Jack Welch, the chairman of General Electric. By 1997, Singh had convinced Welch to set up a business outsourcing unit for GE in Gurgaon. This set a precedent. Soon, other international companies flocked to Gurgaon, and to the offices and apartment buildings that DLF was building.

The present-day showcase of this is Cyber City, a commercial hub created by DLF. Full of gleaming glass and metal buildings, Cyber City houses some of the world's top companies, including Pepsi, Shell, Nokia, Philips, Pfizer, IBM, Hewlett Packard, Ernst & Young, Samsung, American Express, Exxon Mobile, Google, Yahoo!, LinkedIn, and, of course, General Electric. Cyber City was recently connected to the government-run Delhi metro via a smaller metro service run as a joint venture between the government and DLF.

In addition to its role in bringing international companies and IT operations to Gurgaon, DLF has also played a part in transforming the Gurgaon real estate market to make it more amenable to international investment. This has been part of DLF's efforts to expand the National Capital Region and to launch projects throughout India with international backing. This has meant changing the Indian real estate market so that it is more in line with global standards, including the promotion of real estate mutual funds and mortgage markets.[15]

In 2007, when DLF announced its decision to go public, the company looked unstoppable. K. P. Singh had climbed all the way up to number eight on Forbes' well-known list of the world's richest people. After the initial public offering, Singh's wealth skyrocketed from $10 billion to $30 billion. But a nasty surprise was lurking around the corner: the global financial crisis of 2008, which exposed the flaws and risks of DLF's growth strategies.

The immediate cause of the 2008 crisis was the implosion of a sub-section of the American mortgage industry. However, given the increasingly interconnected nature of the global neoliberal economy, the effects of this implosion reverberated around the world. The Gurgaon model depended crucially on international investment, and the crisis left many, including DLF, in the lurch. Prior to the crisis, Lehman Brothers, the firm whose spectacular collapse defined the financial crisis, had invested $200 million in one of DLF's subsidiaries. Merrill Lynch, another company hit hard by the crisis, had invested $370 million in DLF township projects.[16] After the crisis, such investments dried up.[17]

The slowdown after the financial crisis also drew attention to more fundamental issues with DLF's business model. From the beginning, DLF had not been shy about taking on significant amounts of debt to fuel its growth, on the speculative assumption that demand would keep rising and investments would keep flowing in. As a 2015 article on DLF points out, this strategy worked well "in the bull run before the Lehman crisis", when irrational exuberance encouraged market players to take the "aggressive presentations of developers at face value".[18] In the more subdued post-crisis phase, DLF became increasingly saddled with debt.

Reflecting these setbacks, DLF's stock price declined 57 percent between 2007 and 2012. Singh tumbled from 8th place to 191st place on the Forbes' billionaires list. As the company lost its aura of invincibility, the questionable business practices that fueled DLF's rise suddenly started receiving attention. In March 2012, a Canadian firm called Veritas Investment Research issued a scathing report accusing DLF of all sorts of accounting irregularities and business malpractices. The title of the report reflects its exceedingly harsh assessment (as well as its penchant for groan-inducing puns): "A Crumbling Edifice".[19]

DLF Saga

1947   Chaudhary Raghavendra Singh

He buys land from farmers,

and sells to partition refugees.

He develops several colonies in south Delhi and dreams of further expansion,

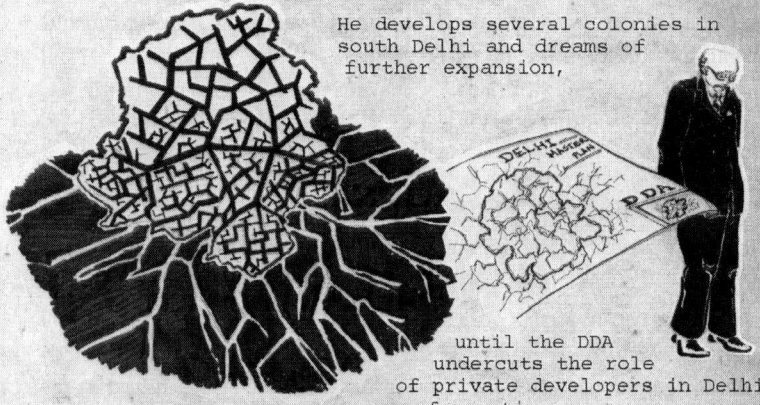

until the DDA undercuts the role of private developers in Delhi ...for a time.

1986   KP Singh takes over DLF. He dreams of building a city in Gurgaon...

and gets the licenses to do so.

He buys land from farmers, and sells to the urban middle class.

In October 2012, reports surfaced that DLF had given interest-free loans and other favors to Robert Vadra, Rajiv Gandhi's son-in-law, who had managed to acquire significant tracts of land at below-market-value prices. Vadra's fixed assets and investments jumped from ₹7.95 crore in 2008 to ₹60.53 crore in 2010. As one news report dryly notes: "It remains unclear why DLF and other major corporations would have made him large loans, since this is not in the nature of their business. Nor did Mr. Vadra's companies have any apparent prior specialisation in real estate business."[20]

Though Congress's political opponents made much of these findings, neither Vadra nor DLF has yet been punished for these shady dealings. Nonetheless, DLF is struggling. As of August 2014, it had accrued ₹19,000 crore of debt, and the Competition Commission of India had brought several cases against it for abusing its dominant position to price its goods and services unfairly. When the Supreme Court gave an interim order on one of these cases, telling DLF to deposit ₹630 crore, the company's stock declined 7.3 percent in two days.

These scandals may help explain DLF's tarnished reputation. More broadly, in the wake of the 2008 crisis, real estate and financial companies have come under increased scrutiny, in contrast to manufacturing firms like Maruti Suzuki. While Maruti fits the stereotypical model of productive economic activity, putting together tangible products, in a big factory, on an assembly line, DLF's work is in the acquisition and sale of land, a pursuit that lends itself to speculation.

As tempting as it is to denounce the "casino capitalism" and rampant speculation of the real estate sector (and, along with it, the financial sector), the "productive" economy could not exist without the land and the liquidity that these sectors provide. This is one of capitalism's many contradictions: while these sectors are essential for capitalist growth to function smoothly, they also enable all kinds of speculation and distortions. Financial and real estate investments are, at their core, a gamble; they are speculating on what future returns will be. In Marx's evocative language, these sectors traffic in "fictitious capital", which acts as the savior of accumulation, but at the same time "the fountainhead of all manner of insane forms".[21] The DLF model for Gurgaon shows the risks of such insanity.

Besides its failure in purely economic terms, the urban model promoted by DLF has been an ecological disaster. Runaway economic growth has also meant rapid population increases. According to one estimate, Gurgaon's population is growing at a rate of 250 percent per decade. Gurgaon has started to recognize its error in not heeding the Delhi Master Plan's ominous warning about lack of water resources in the area. One dire article has proclaimed that Gurgaon is "on its deathbed",[22] as the city obliviously extracts the last of its groundwater. This problem has been exacerbated by the destruction of traditional bunds, which channeled water from the Aravalli hills and kept the city's small lakes and ponds alive.

One prominent example of this is the construction of a Maruti Suzuki manufacturing plant in Manesar, a village bordering Gurgaon and now part of the same industrial zone. The plant has blocked the flow of water coming down from the Aravallis, which has led to waterlogging on one side and dessication on the other. There was no due diligence completed before the construction of the plant, and no consideration of its ecological impact. Despite their differences, DLF and Maruti Suzuki are bound together in the same destructive model of development.

Delhi, too, a "World-Class City"

Since the 1980s, then, the National Capital Region has seen profound, and deeply unsettling, economic and environmental changes. Gurgaon is often presented as the poster child of these changes, but Delhi too has seen radical transformations. Though Gurgaon is technically in a separate state, the growth of Delhi and the growth of Gurgaon have always been intertwined. Gurgaon, after all, was marketed as a suburb of Delhi (its proximity to Delhi's international airport has long been a selling point), and it has been connected to it by the Delhi Metro since 2010.

The Metro is a telling manifestation of the rapidly diminishing distance between Delhi and Gurgaon, both practically (in terms of travel time) and geographically (in terms of physical distance between the urban centers). The Metro ride from Delhi to Gurgaon is fascinating. After the hubbub of Chhattarpur station, near Mehrauli, the scenery slowly begins to change. Buildings start to thin. Farmhouses come into view, the elevated rail line disturbing the jealously guarded privacy of the estates' owners. The landscape is dotted with wedding halls and swimming pools, surrounded by expansive green lawns. After the farmhouse belt, a profusion of trees appears, a sea of *vilayati kikar* covering the Reserved Forest section of the Southern Ridge. And then, abruptly, the greenness ends, and the rider is deposited in Gurgaon, amidst the malls and the dystopian, futuristic steel and glass structures.

The Reserved Forest appears as an oasis, a calm refuge from the twin urban centers now surrounding it. But the peaceful appearance is deceptive, as this zone, along with the farmhouse belt around it, has been the site of intense contestations, both ideological and physical, between different land uses, and the different people who champion those uses.

In this respect, it is a mirror of the city as a whole. Delhi, in its Ridge and beyond, has followed Gurgaon's lead in imagining itself as a "world-class city", with all the requisite, globally recognized symbols of conspicuous consumption: a proliferation of malls and multiplexes, store after store of international brands. But the re-imagining of Delhi has not just been confined to its sites of consumption; economic production has also gotten a makeover. Mines on the Ridge have closed, as we saw in Chapter 1, but this is just one sign of a broader shift: the DLF-ification of Delhi, with real estate, international investment and IT taking the place of old-fashioned manufacturing, at least in the dominant imagination of the city.

It was largely a matter of imagination: in reality, both before and after liberalization, Delhi's economy has been dominated, not by manufacturing, nor the financial sector, but by trade and services. This reality, however, did not stop the many re-imaginings of Delhi as a "world-class city", which had as their model, whether implicitly or explicitly, the post-industrial, consumerist Western city.[23] And this imagination had a very real impact on Delhi and its Ridge.

In many Western cities, industry left of its own accord, as cheaper labor markets emerged abroad. In cities like Delhi, the process was quite different, as the courts and a section of Delhi's citizens conspired to push industry out of the city. The most dramatic case of this was the M. C. Mehta PIL, discussed at length in the previous chapter, which led the Supreme Court to order the closure of thousands of industrial units, and which later spurred the expansion of the Ridge's Reserved Forests.

While Mehta's motivation was clearly environmental, his PIL got caught up in larger economic and aesthetic transformations of the city. The Supreme Court explicitly acknowledged that factory land would be more profitable if it was converted from industrial use to commercial, residential and office use.

Reviewing the court order mandating the closure of industries, it becomes difficult to disentangle economic and environmental motivations. The order stipulates that a third of each industrial plot should remain as the private property of the owner, but that the other two-thirds should be given to the DDA to create open spaces, green areas, and lung spaces, the language of Jagmohan returning in full force. Despite their shrunken landholdings, the owners should be grateful for this court order, since the beautification of the area would push up property values, and hence the owners should "see a gold mine in them".[24] Clearly, the court was using this term in a purely metaphorical sense, since mining had also recently been banned in Delhi, on very similar grounds, which had helped ignite the farmhouse phenomenon, another sign of the increasing role of speculative real estate in Delhi's economy.

A slew of court orders and government policies like this cleared the space for land uses that seemed appropriately modern, from offices performing Business Process Outsourcing (BPO) services to farmhouses hosting lavish wedding parties. Along with industries and mines, slums were a chief target of Delhi's neoliberal transformation. The DDA now explicitly defines a "world-class city" as a "slum-free city".[25] The courts also drove this message home, while stressing Delhi's heritage as a capital, a trope that should be familiar from the previous chapter. One anti-slum court ruling mused,

> Delhi being the capital city of the country, is a show window to the world of our culture, heritage, traditions and way of life. A city like Delhi must act as a catalyst for building modern India. It cannot be allowed to degenerate and decay.[26]

The rhetoric of a "world-class city" has two dimensions. The first is a desire for the world to see Delhi in a certain way: as slum-free, as modern, as developed. The second is a reconceptualization of Delhi's relationship to its surroundings. Though its role as national capital is still emphasized, the city is increasingly seen, not as a hub for its rural hinterland, but as one node among many in a global economy connected through the magic of the IT revolution. These two aspects of the "world class city" are closely linked: it is through the projection of a "world-class" image that political and business leaders position the city as an appropriate site of international investment. Both aspects show a marked preference for style over substance: the important thing is what the city looks like on the surface, not the exploitation and violence and chaos beneath that surface.

This chaos can be seen in informal settlements like the one in the shadow of Qutb Minar, described at the close of the previous chapter. Such scenes of demolition and painful regrowth have proliferated across the city. These demolitions are particularly galling because they do nothing to address the root causes of the housing crisis in Delhi. Settlements are destroyed, only to arise again due to the lack of affordable housing and, in recent years, a set of government policies that have explicitly encouraged the growth of cities. The post-1991 era in Delhi has seen a flood of migration into the city, with rich and poor alike streaming in. Between the censuses of 1991 and 2011, the official population of Delhi rose by seven million, and this is almost certainly an underestimate.

This increased migration is in large part due to the agricultural crises facing Delhi's hinterland in particular and the country's rural areas more generally, which have largely been neglected in the projection of a "New India", and which find their most tragic manifestation in the ever-increasing number of farmer suicides across the country. In many ways, working-class urban neighborhoods, including the many that dot the Ridge, are living histories of rural displacement. Poor migrants have been swallowed up by a city ravenous for workers to mine its quarries, to build its malls, to construct its roads and buildings, and to clear the way for a new infrastructure of internet cables and satellite TVs. They are absolutely necessary for the creation of an appropriately world-class Delhi, but they fall into the shadows of the city, seen as an affront to moneyed property owners when at home and relegated to the shady, unregulated world of subcontractors, middlemen and informality when at work.

Nouveau Riche on the Southern Ridge

The past several decades have also brought to Delhi those at the upper echelons of these new value chains of informality and illegality, masters of black money circuits, political influence and back-room negotiations. The professional, salaried middle classes resent this new bourgeoisie, who are operating by another, wilder set of rules. These intruders are seen as uncouth and uncultured, as the nouveau riche usually are, but their wealth is also envied. Their lives are shrouded in secrecy, but two recent high-profile murders on the Southern Ridge have offered a rare glimpse into the world of new money in Delhi. The media coverage of these cases has shown the public the sheer wealth of Delhi's new elite, in contrast to the crushing poverty that marks the lives of the city's majority. But the grisly details of these stories also show the volatility of these new fortunes.

It is no coincidence that these cases center on the Southern Ridge, and specifically on Ridge farmhouses. By calling their luxury estates "farmhouses", Delhi's elite can claim the water subsidies available to Master Plan-zoned agricultural land, while enjoying the combination of relative seclusion from and proximity to South Delhi and Gurgaon. In the memorable words of Dilip Bobb, the farmhouses "display a bewildering range of architectural indulgences, from Spanish villas, to American-style ranch houses and marbled mansions that would give Beverly Hills an inferiority complex".[27] Following in the footsteps of Metcalfe, the nouveau riche of Delhi have used their property on the Ridge as prominent status symbols. But like Metcalfe, two new claimants to Delhi's prestige have met violent ends, felled by a mix of political maneuvering and family feuds.

The most famous case is that of Ponty Chadha, the victim of a highly publicized farmhouse shootout on the Southern Ridge. Chadha presided over a diversified business empire, though he was best known as the "liquor baron" of Uttar Pradesh. In November 2012, Chadha was killed in a confrontation with his brother outside the family's sprawling DLF Farms mansion. But Chadha was a late-comer to Delhi, and for him, the Ridge farmhouse was merely a status symbol. The next high-profile farmhouse murder was different. Here, the victim was Deepak Bhardwaj, who made his fortune in Delhi and precisely through its real estate. For him, the Ridge was not just a place to display his wealth; it was his means of generating wealth as well.

Bhardwaj once entered the Lok Sabha elections, but was trounced. His real success had come not from electoral politics, but from land deals, with a focus on the Master Plan's Green Belt, including many properties on the Southern Ridge. The farmhouse in which he met his demise was his residence, but it was also the site of a massive "hospitality hub" called "The Nitesh", named after Bhardwaj's younger son, who was married to Shah Rukh Khan's sister-in-law for a time. The venue hosts events, conferences, parties and (inevitably) weddings.

The website of "The Nitesh" details the facilities available: a 28-acre, four-lawn complex; acres of lush greens, shady trees and flowers; exclusive in and out gates for each lawn; banquet hall space of over 35,000 square feet; a 40-room hotel; well-appointed bridal rooms and toilets; golf carts to ferry guests. Each lawn is described in detail, with a focus on features like "premium wooden furniture", "large granite porch", "*vedi* enclosed in a glass pyramid", and "funky props and accessories". People familiar with the farmhouse/wedding industry have estimated that the venue brought in a monthly revenue of roughly ₹2.5 crore.[28]

In the days following the murder, revelations about the case mounted, taking on a stranger-than-fiction hue. Subtlety was not the preferred method of the killers. They drove straight to the front gate of The Nitesh and asked to meet Bhardwaj. After going inside, they shot Bhardwaj point-blank with country-made pistols. They then ran out of the complex, guns drawn, and jumped into their getaway car.

The assailants were quickly apprehended by the police. But it was clear that they were only following orders. Early news reports indicated that the killers had gotten their instructions from Swami Pratimanand, a self-styled godman from Haridwar. Pratimanand wanted to establish his own ashram but did not have enough money.

He even attended the massive Kumbh Mela earlier in the year in order to impress akhara gurus. However, other holy men did not take him seriously because he did not remain at any one ashram for more than a year. Restless, he was seeking ways to make a quick buck. He was told that he would need at least ₹1.5 crore to create an ashram that would attract foreign tourists. The contract for Bhardwaj's killing was supposedly ₹5 crore.

Pratimanand, then, was not the main mastermind of the killing. He too was just a hired hand. It was soon revealed that Pratimanand was working for a lawyer and real estate agent named Baljeet Sehrawat, who nursed a grudge against Bhardwaj because of a real estate deal gone sour. Sehrawat had helped Bhardwaj with a disputed property deal, but after the deal was done, Bhardwaj refused to pay Sehrawat's commission and legal expenses. Sehrawat can be seen as a younger, more ruthless, version of the man he had killed: an up-and-coming real estate mogul with political ambitions.

But the most shocking revelation was yet to come: Sehrawat was not the main conspirator in the case. He too was just following orders. As the case unraveled, the true culprit was finally found. The hit had been ordered by Bhardwaj's younger son, the very Nitesh after whom the family estate is named.

The cause of the family drama was allegedly a women named Sonia, who had become Bhardwaj's lover. Thirty-three years Bhardwaj's junior, Sonia had become an integral part of The Nitesh's business operations. This did not sit well with the family, invariably described in media reports as estranged from Bhardwaj. According to the family, Bhardwaj was withholding money from them and keeping them in conditions of "near poverty". The children feared that he would disinherit them and give all his money to Sonia. While Sonia accused the family of ordering the hit, the family in turn claimed that it was actually Sonia behind the murder. The police were more convinced by Sonia's version of events; they arrested Nitesh. The police also claimed, in a bit of tragic irony, that Nitesh used his father's own money to pay for the contract on his life.[29]

In this murky intersection of crime, real estate speculation, religion, love and family, it is difficult to determine the true motive for the killing. Every accusation meets with a counter-accusation, and one can hardly trust the leaked reports of the Delhi police. However, it is quite possible to see the larger developments that led to the proliferation of figures like Bhardwaj, a low-level government stenographer who became the richest political candidate in Delhi's history and who owned a prized farmhouse on the Southern Ridge.

Bhardwaj himself described his rise simply and bluntly, "Real estate is the surest way of getting rich."[30] His death led to extensive reporting on his life, from his early childhood to his untimely demise. The picture of Bhardwaj that emerges is not a pretty one.[31] He was, it appears, singularly obsessed with wealth, and with the acquisition of real estate as a means of amassing wealth.

Bhardwaj was born in the village of Chetiya Oliya in Haryana. His father worked as a carpenter, and the family grew up poor. After completing a bachelor's degree in commerce from Delhi University's School of Correspondence, Bhardwaj moved to Delhi and became a stenographer for the city government's sales tax office. His office was located in the Tis Hazari Court complex, and here Bhardwaj got a thorough education in the intricacies of land deals, including ways to exert legal pressure and corner valuable pieces of real estate. He began to save up money to invest in land.

When he was tired of the bureaucratic life, he decided to open an automobile parts business so that he could take advantage of the rising sales of scooters and cars—these were the early days of Maruti Suzuki's success. But he soon realized that there were more lucrative occupations in the quickly expanding city. He became involved in real estate deals, first part-time and then full-time. He specialized in buying agricultural land and then either selling it as farmhouses or exerting political pressure so that the land use was converted to "industrial". He made massive amounts of money when the government acquired some of his land for the expansion of the city's airport, and he re-invested the money in various rural properties that he thought would soon be urbanized.

Bhardwaj was one of hundreds of brokers working on the rural edges of Delhi, away from the monopoly of the DDA, and away from DLF's growing Gurgaon empire. He, like many others, benefited from the DDA's failures to build enough housing in the "urbanized" part of Delhi, which drove potential homeowners to so-called "agricultural" areas. Bhardwaj also took advantage of the breakdown of the village commons system. By exploiting various loopholes, powerful villagers took private ownership of many "common" lands, just as had happened in the villages surrounding Mangarbani (as detailed in Chapter 2). They then sold these lands to real estate dealers, even though the Master Plan technically prohibited this. A savvy middleman, Bhardwaj allied himself with influential *panchayat* members in various villages, buying up land himself or connecting villagers with wealthy buyers.

Business associates remember his ability to sweet-talk potential business partners and his command of English. But he was also able to relate to villagers who were just coming to terms with, and hoping to profit from, the coming urban onslaught. During his early days in the Delhi real estate business, he would ride around the hinterland on a two-wheeler, wearing a white kurta-pajama, with a pistol strung around his neck. The expansion of Delhi, and the conversion of barren Ridge land into high-end real estate, depended on middlemen like Bhardwaj, who were able to connect with both the villagers selling their land and the Delhi elite buying it.

Bhardwaj particularly focused on properties that were under legal dispute, stuck between the strictures of the Master Plan and the rather messier realities on the ground. He would sell these properties to third parties, knowing that this would almost certainly draw lawsuits. He had hundreds of civil cases pending against him, but from his time in Tis Hazari, he knew that such cases often dragged on endlessly, and that he could make enormous amounts of money while they did. Unlike most real estate moguls, he was happy to appear in court and personally attended all his hearings. Backed by a team of lawyers and with strong connections in the government, Bhardwaj was able to turn huge profits on disputed lands. Bit by bit, he built up a massive real estate portfolio.

While Bhardwaj had been quietly amassing wealth for decades, the media spotlight first shone on him in 2009, when he decided to contest Lok Sabha elections. He ran as a candidate for the Bahujan Samaj Party (BSP), and, in keeping with government regulations, he had to declare his assets: ₹600 crore. A vast majority of this was from "agricultural" land: 368 crore owned by Bhardwaj himself, and 100 crore in his wife's name. After losing the election, Bhardwaj was reportedly building his connections with BJP leaders, and, just before his death, he hosted a *yagna* in honor of Narendra Modi.

Circling Back to the Vasant Kunj Ridge

The Bhardwaj and Chadha scandals brought out the moralistic side of Delhi's middle class. Though representatives of this class were often the loudest cheerleaders for neoliberal reform, many have recoiled at the "uncivilized" exploits of the nouveau riche who populate the wild post-1991 economic landscape. In the comments section of an online *Times of India* article about Chadha, the following remarks could be found: "How much ever wealth you amass, legally or illegally, you need to have certain amount of education and upbringing to sustain or grow it, which Ponty & his brothers lacked" and "Money without education is like a weapon of mass destruction".[32]

As these comments suggest, middle-class resentment is equally channeled towards the unattainable farmhouses and the sprawling slums. In the mainstream, middle-class environmental discourse—the one to be found in the pages of the English daily newspapers—both slums and farmhouses are seen as encroachments that the government should remove. This apparent even-handedness, though, does not take into account the actual ways that power works in Delhi, the larger political and economic networks shielding Delhi's new moneymakers from punishment no matter how brazen their illegalities. In practical terms, this means that poor "encroachments" are demolished, while rich ones are left unscathed.

This point was brought home forcefully during the decade-long struggle over the fate of the Vasant Kunj Ridge, where DLF and Maruti Suzuki proudly stand together. In some sense, this is an old story: the state is using the Ridge as a prize, as a site to display prestige and glory. But the story has a distinctly neoliberal twist: the state was flanked by private actors in an effort to transform the Ridge into a symbol of Delhi's new grandeur.

The seeds of the controversy were planted in 1994, though few recognized it at the time. When the tentative boundaries of the Ridge (as Reserved Forest) were drawn in 1994, they were done using maps provided by the DDA. According to the physicist and environmental activist Vikram Soni, who became one of the chief advocates for maintaining the Vasant Kunj Ridge as a forest, the DDA intentionally left some Ridge zones off the maps so that they would not become protected areas.

To activists working on this case, it hardly seemed accidental that many of the missing Ridge zones were in South Delhi, where real estate prices were continuing their ascent, and where speculation was rampant. The DDA was a main player in this speculation and seemed intent on "developing" Ridge land, not with forests or parks, but with more lucrative ventures. A main area of interest was the plot of Ridge land that lay between the DDA-approved neighborhoods of Vasant Vihar and Vasant Kunj.

The DDA's first move in this area was fairly minor. In 1995, the government started to build a road between Vasant Vihar and Vasant Kunj, cutting through Ridge land. Soni, however, found this alarming, as he had been taking refuge in the quiet, wooded solitude of the Ridge since moving to Vasant Vihar five years earlier. He recognized that a road was just a warning shot in a larger battle over the land.[33]

The Vasant Kunj Ridge was, at the time, an accidental forest. It had been used for quarrying until the 1980s, when mine workers finally dug all the way down to groundwater levels and mining stopped. The abandoned, denuded landscape did not stay barren for long. Within a few years, ponds had formed in the quarry pits. Seeds had blown in as well, lodging in cracks and crevices. Soon enough, trees, shrubs and grasses started to grow. This created a dramatic, and increasingly green, landscape of hills, valleys, trees and lakes. Here is one more reminder that the Ridge is hardly a land of pristine glory. The corollary to this lesson is that even drastic human intervention like quarrying need not spell the death of the Ridge, and can even provide new micro-habitats, if the land is given the space and time to regenerate.

Soni had watched this regeneration first-hand, and he was not ready to see the land fall into the DDA's hands. He knew that Delhi had passed a law outlawing the felling of trees without special permission.

He asked the road builders if they had obtained permiss   d he was met with stony silence. After some more research, he   riend filed a PIL with the Delhi High Court and succeeded i   ng the construction of the road.

Their victory was short-lived. In 1996, even as the Delhi government was taking away villagers' common lands to make Reserved Forests, the DDA announced that it would develop a luxury hotel complex on the Vasant Kunj Ridge, which would house 13 five-star hotels. This reveals much about the government's imagination of a clean, green city: a densely-packed cluster of high-end hotels, which would consume massive amounts of water and electricity, could profitably be placed on the Ridge, as long as other parts of the Ridge were set aside for forestry (after being wrested from the villagers that owned these lands).

Soni immediately began to campaign against the hotel complex. He had an influential ally: Kuldip Nayar, also a Vasant Vihar resident, a journalist known for his unbending integrity. He was one of the few journalists who stood up to Indira Gandhi during the Emergency, and he was jailed for it. Soni and Nayar started going on walks together through the wilderness of the Vasant Kunj Ridge, and Nayar soon joined the fight against the DDA and the hotels.

They were accompanied by others in Vasant Vihar who did not, perhaps, share Soni and Nayar's ecological appreciation of the Ridge, but had other concerns about the hotel project. Many Vasant Vihar residents worried that the hotels would lead to the creation of a busy thoroughfare connecting their neighborhood to Vasant Kunj, which they saw as decidedly down-market. This was, of course, a subjective judgment. The DDA flats at Vasant Kunj were highly coveted by most of the middle class; compared to the posh Vasant Kunj, though, they lacked a certain glamor. Soni is quite frank about his neighbors' motivation, referring to this faction in jest as the "elite gentry of Vasant Vihar".[34] He may have questioned their attitude, but he certainly welcomed their support.

The DDA, meanwhile, had its own considerations in mind. A luxury hotel complex was very much in keeping with the new, "world-class" image that the DDA wished to promote. It would draw high-end international tourism and make the city attractive to international businesses. The Vasant Kunj Ridge was close to both the airport and Gurgaon; it was thus the perfect location for those wanting a global feel.

Again, Soni and his allies took their case to the court, and again, they found a sympathetic judiciary. While the Vasant Kunj Ridge was not included in the 1994 Reserved Forest notification, the Vasant Vihar residents still believed that the law was on their side. They pointed to the second Master Plan for Delhi (published in 1990, and meant to guide development in the city until 2001), which defined the Ridge geologically, as the "rocky outcrop of Aravalli hills", and mandated that "no further infringement" be allowed. The Vasant Kunj Ridge clearly met this definition. The Supreme Court agreed with this logic and ordered a stay on the project. The court also mandated the creation of the Environment Impact Assessment Authority (EIAA) to review this project and others like it. By this point, land had already been cleared for the foundation of a hotel called the Hyatt Grand, a collaboration between Hyatt, the American hotel giant, and an Indian company called Unison. After the court order, construction was halted.

But the DDA was patient, as was the Unison management. Kuldip Singh, the judge who had pushed for the creation of the EIAA, and who was most sympathetic to the petitioners, retired in 1997. After Singh's retirement, Unison kept on filing petitions in court, trying different strategies to convince the judiciary to allow construction to resume. By 1997, they had found the right formula. Advancing a very questionable reading of the DDA's initial plans and previous court findings, Unison's lawyers insisted that the DDA's plan only referred to 223 hectares of Ridge land, in a larger plot of 315 hectares. So when courts had earlier ruled that the hotels could not be built, they were only referring to the 223-hectare plot; however, since the Hyatt hotel was located on the smaller, 92-hectare plot, its construction should be allowed. This clever argument ignored the fact that the EIAA had quite clearly stated that all 315 hectares should be considered protected. Nevertheless, the argument was accepted by the court.

For this case, the hotel's main lawyer was Harish Salve, a well-known, well-connected senior advocate. But he got support in his creative interpretation of the law from the advocate representing the DDA, Arun Jaitley, another legal and political heavyweight (most prominently, he served as the Minister of Finance from 2014 to 2019). Soni, Nayar, and others had put their faith in the legal system, figuring that they had an airtight case. But powerful forces and influential figures, both in the government and in the private sector, knew that, especially in the city of Delhi, the law is flexible, and can be made to bend in all kinds of surprising contortions.

The construction of the Hyatt Grand thus resumed, and it now stands on one corner of the Vasant Kunj Ridge. The hotel's website emphasizes its luxury ("a glamorous wonder") and its international pedigree. Its exterior was designed by a London-based firm, its interior by a San Francisco-based one. The website also has the temerity to note that the hotel has a "fine location near the [city's] dense green belt".

Round 2

Those fighting the DDA's plans could console themselves with the fact that, due to their efforts, only one five-star monstrosity was built, not the whole 13-hotel complex. But the DDA knew it had stumbled on a successful formula. The courts had admitted that the 92-hectare plot fell outside of the zone where construction was prohibited. The Grand Hyatt (now, with a new ownership arrangement, simply called "The Grand") only occupied four hectares of this plot —what of the other 88?

The DDA soon answered this question. In late 2003, it started advertising an auction for plots of land on the Vasant Kunj Ridge, to be used for the development of malls and corporate offices. This would take up 25 of the 88 hectares, and, from the beginning, was projected as a prestigious, luxurious new development. Two of the auction's eventual winners were quite recognizable: ONGC, a public-sector company and the biggest oil and gas company in the country, which sought to build a new corporate headquarters on the plot, and Ambience, a well-known mall developer. The other winners did not seem as notable, at least to a casual observer. They included companies called Beverly Park Maintenance Services, Regency Park Property Management Services and Jasmine Projects.

But these names concealed the presence of bigger players, as opponents of the malls soon found out. Beverly Park Maintenance Services and Regency Park Property Management Services were both subsidiaries created by DLF to pursue various construction and development projects. The malls were eventually branded with the DLF tag and became DLF Promenade and DLF Emporio. Jasmine Projects was a real estate developer that had teamed up with Maruti Suzuki.

These were powerful players at the height of their Gurgaon-based success, eager to expand in the capital. In the ensuing legal battle over the commercial complex, they were backed by the same high-power lawyers, but with a slight twist. This time, Harish Salve, who had earlier represented the hotels, now served as the DDA's advocate. Meanwhile, Arun Jaitley left his responsibilities with the DDA to represent the malls and corporate offices. This neat switch was a clear indication of the new political and economic situation in Delhi: in the eyes of the project's backers, the interests of the state and the interests of private companies were identical.

Those fighting against the malls and offices thus had their work cut out for them. An early legal challenge against the proposed commercial complex was dismissed, and construction began in September 2004. But the movement against the project had gained steam, attracting the attention not just of Vasant Kunj residents, but of environmentalists and NGOs across the city. The protest movement also had the backing of some sections of the government. The environmentalists fighting the case got an unexpected boost from the Delhi Pollution Control Committee, which asserted that the malls and offices had not submitted proper Environmental Impact Assessments (EIAs), and ordered a halt to the construction. Further, the Central Empowered Committee (CEC) of the Forest bench of the Supreme Court found that the plot in question was definitely Ridge land, and was a valuable water recharge zone as well, and thus deserved protection as a Reserved Forest.

As lawyers for both sides presented their case to judges and to government committees, another battle was taking place in the court of public opinion. Environmentalists organized protests and drew on their contacts in the media to launch a sustained campaign against the commercial complex. They also showed up in large numbers to public hearings organized by the government about the controversy. But their opponents also came prepared for these meetings. For the largest, most contentious public meeting, in the summer of 2006, the malls' backers had brought in hired muscle to crowd out the protesters and drown out their voices.[35]

Soon after this meeting, an Expert Committee from the Ministry of Environment submitted its report on the case. It started the report by reiterating what environmentalists had been arguing for nearly a decade:

> *Various studies, including EIA documents submitted now for obtaining environmental clearance, establish the environmental value of this area.... Therefore, DDA should have exercised adequate environmental precaution based on sustainable environmental management approach. There is no evidence that the environmental impact of the construction of malls was assessed beforehand.*

But after stating all this, the report took an unexpected turn:

> *In hindsight it is evident that the location of large commercial complexes in this area was environmentally unsound. Now many proponents have constructed very substantially and really speaking awarding clearances even with conditions is largely a compromise with de-facto situation. The Expert Committee is of the opinion that at this stage only damage control is possible.*[36]

In essence: the project is an ecological disaster, but construction has already started, so we can't stop it now.

For the environmentalists working on this case, the Ministry's about-face was shocking. Construction had, indeed, started, but hardly in a "very substantial" way. The foundations of the buildings had barely been laid. "Very substantial" was the clout of the project's backers, and the prestige, money and power they had brought to the table. In the end, the Supreme Court accepted the recommendations of the Expert Committee. In its deliberations, the court also noted the value of the malls in bringing international brands to the city.

The "damage control" referred to by the Expert Committee was also hard to identify. The malls and office were built with no major changes to their initial plans after the companies involved had submitted perfunctory EIAs. As punishment for their environmental crimes, the companies were fined a mere ₹5 lakh, this on a plot of land that real estate insiders valued at ₹25,000 crore.

Though attention was focused on the malls during the legal battles that took place between 2003 and 2006, parallel processes were unfolding behind the mall construction site, on the remaining portion of the Vasant Kunj Ridge. In fact, the malls and offices would only take up 25 hectares of a plot of Ridge land that totaled 640 hectares. Of this, 315 hectares were controlled by the DDA and 325 hectares were controlled by the Army. In the early 2000s, the Army began massive construction efforts on the Ridge, cutting down trees and laying foundations for schools and housing complexes. Soni and his allies tried to bring media attention to this case, with mixed success, and he even arranged meetings with high-level Army representatives. Though they resented the bad press, Army officials realized that other branches of the state would not really put a stop to their construction efforts. At worst, they would impose small fines. Their building projects thus continued.

The DDA, in the meantime, contemplated what it would do with the remaining 290 hectares of its Vasant Kunj Ridge land. 10 hectares were used to establish a small institutional area, which would host, among other things, a university established by The Energy and Resource Institute (TERI). The irony of this was not lost on environmental activists; TERI presented itself as India's leading environmental NGO (its founder, Dr. R. K. Pachauri, was chairman of the Intergovernmental Panel on Climate Change when that body won the Nobel Peace Prize), and yet it was gobbling up land on Delhi's Ridge. When confronted by activists, TERI officials meekly replied that the DDA had assured them it was not Ridge land.

This still left 280 hectares of land. Officials recognized that the DDA was getting pilloried in the press for its callous attitude towards the environment. Their solution was to convert the remaining portion of the Vasant Kunj Ridge into a biodiversity park. This was the zone that had been mined most intensively; it would have been difficult to construct either malls or Army buildings on such a pockmarked land-scape. This area did, though, house a small working-class settlement. It was promptly demolished, with little media fanfare.

Despite the dubious roots of the park's establishment, the DDA showed a genuine interest in the biodiversity project once it began in 2004. (This is in marked contrast to the early years of the Asola Bhatti wildlife sanctuary, when *vilayati kikar* proliferated, mining continued, and trucks barreled through the "protected" zones.) DDA officials sought the help of ecologists at the Centre for Environmental Management of Degraded Ecosystems (CEMDE) at Delhi University, who made careful forestation plans for different zones of the park. The park now hosts guided nature walks and even has a small campsite for visiting students.

Despite these outreach activities, the Biodiversity Park is largely hidden from public view. In part, this is by design; the ecologists who have spent over a decade designing the park are wary of visitors who may litter, trample on plants, and generally wreak havoc. But the invisibility of the park is largely determined by its surroundings. It is sandwiched between, and dwarfed by, luxury malls and Army settlements. These two complexes hide the park in different ways. The Army, shunning media attention and valuing tight security, makes it difficult to approach the Vasant Kunj Ridge from the west side. The malls are more porous (although security guards are present to turn away those who look like they can't afford the goods offered inside), but they draw all the attention to themselves. Style over substance: the malls set the tone for the Vasant Kunj Ridge.

Why Malls?

The Vasant Kunj malls are among Delhi's biggest and most luxurious. They are a potent symbol of the city's economic transformation in the age of globalization. They underscore the economic shift from the manufacturing sector to the service sector. But more importantly, they signify a new kind of consumerism and a new image of India's gleaming future, a balance of global cosmopolitanism and Indian tradition. In short, the malls bring together the generation of surplus and the conspicuous consumption of surplus on the Ridge.

This is not what malls were initially intended to do. The shopping mall's inventor, Victor Gruen, was a fervent socialist. Born in Austria, Gruen moved to the United States in 1938. In a 1952 article for the journal *Progressive Architecture*, Gruen outlined his vision for the mall: it would be a center of social life, not just for shopping, but for living, working and socializing. Densely packed stores would be surrounded by apartments, office complexes, hospitals and schools. Gruen saw this as a remedy to suburban sprawl and as a way to create vibrant, integrated meeting places. Gruen lived long enough to see his dreams dashed, and to watch with dismay as malls only accelerated sprawl and became isolated realms of intense consumption. Towards the end of his life, he repudiated the entire idea of the mall, saying, "I refuse to pay alimony for those bastard developments."[37]

One of the reasons his ideas failed is that it is exceedingly difficult to build a socialist space in the midst of a capitalist economy. More specifically, the real estate market, so central to the story of contemporary Delhi and the Ridge, had functioned in similar ways halfway across the world, encouraging speculation and foiling the best-laid plans of urbanists and architects. Gruen's first mall project, in the midwestern state of Minnesota, envisioned the mall as the centerpiece of a multi-use 500-acre development, where people could live, work, shop, eat and play. However, as soon as the mall itself proved to be successful (75,000 people visited on its opening day), property values around the mall skyrocketed. This made it difficult to build schools, hospitals or parks in the vicinity; instead, the land was snapped up by other commercial enterprises, as well as by high-end housing developers, who proceeded to worsen the very sprawl Gruen was seeking to contain.

After this, malls rapidly spread across the United States and became the signature feature of American suburbia. This, in part, explains their draw in Delhi; they are a symbol of the 'developed' world that many want to emulate. But ironically, by the time malls started to flourish in India, they had begun their precipitous decline in the US. By 2012, about one-third of all American malls were either totally defunct or on the verge of dying, as they struggled with low footfalls and the exodus of high-end brands who had started to see malls as passe. In one case, a mall in Columbus, Ohio was torn down and replaced by a park, an inversion of the Vasant Kunj Ridge process.[38]

Malls came to India in the 1990s. Depending on who you ask, the first mall in India was either in Mumbai or Chennai, but Delhi soon became the prime location for new malls. The political capital of the country was increasingly remade as its consumption capital. This trend extended to the larger National Capital Region; one stretch of road in Gurgaon features a staggering 11 malls, with more in the works.

Malls were attractive for their symbolic association with the West, but also for more mundane, number-crunching reasons: study after study has shown that malls offer real estate developers the highest rates of return (at least in countries where malls are not dying). Companies like DLF were eager to cash in on this trend, so much so that, in 2009, retailers in DLF Emporio went on strike to protest the high rents that DLF was charging. These luxury-brand managers claimed they were being treated like "bonded labor", a comparison that real bonded laborers (like the ones once employed in the Ridge's quarries) would likely find quite distasteful.[39] Despite the dramatic rhetoric, though, the retailers and DLF reached an agreement about rental prices, and the strike was called off.

Like the IT sector, malls have been projected as a key part of India's entrance onto the world stage. But like IT, malls play a relatively minor role in the Indian economy. Their importance is largely symbolic. In economic terms, malls are considered part of the "modern retail" or "organized retail" sector. These seemingly neutral terms are in fact quite loaded, suggesting as they do that the bazaars and corner shops that dominate the retail market are quaint, old-fashioned, and in a state of disorganized disarray.

Boosters of the new economy have salivated at the prospect of "modern" retail's rise. One early study predicted that, by 2013, more than three-quarters of all retail space in the Delhi region would be dedicated to malls. This has not come to pass, in part because the financial crisis of 2008 disrupted many of the grander plans for mall construction. In 2011, "modern" retail made up about 10 percent of the total retail business in India, although even this may be an over-estimate, due to the difficulty of accurately calculating cash flows in the informal economy.[40]

Looking at the Vasant Kunj commercial complex, it is easy to see why malls have impacted the popular imagination in ways that are disproportionate to their actual economic weight. For one, the Vasant Kunj development is huge. It dominates the landscape. The three malls, Ambience, DLF Promenade and DLF Emporio, are located in the center of the complex, flanked by the Maruti Suzuki corporate office (now also shared by Bharti Airtel) on one side, and the ONGC and ONGC Videsh offices on the other.

The malls, and particularly the centerpiece, DLF Promenade, are built in a style that oddly recalls the domed, colonial-era buildings of Lutyens Delhi, with neoclassical grandness and the occasional, vaguely Indian motif. DLF Promenade is a Viceroy's Lodge for the neoliberal age. Both, not incidentally, are sited on the Ridge, and both were heavily promoted by the state. For Raisina Hill, the state employed private contractors, but of course retained ownership of the site. For the Vasant Kunj Ridge, the state handed the land over to its former enemy, DLF, its longtime friend, Maruti Suzuki, and several others.

It's not just the majestic exteriors of the Vasant Kunj malls that announce their significance. The interiors are also grand, with columns, chandeliers, and massive, high-ceilinged atriums. Nehru once described dams as postcolonial India's new temples; perhaps the malls are the new temples of 'new India'. Certainly, malls like those in Vasant Kunj have been built to evoke a sense of wonder.

The interiors offer more worldly pleasures as well: the opportunity, not just to shop, but to see and be seen, to socialize, to gossip, to escape the heat and grime of the city. Although the malls sell themselves as Delhi's most luxurious, they actually cater to a range of customers, although not, of course, to the poor. They are arranged in a clear hierarchy: the relatively modest Ambience Mall on one end, the posh Promenade in the middle, and the exclusive Emporio at the other end. All the malls offer a mix of Indian and international brands, but the prestige of the brands increases incrementally. The first, for instance, houses Big Bazaar and McDonald's; the second features a boutique called Kama Ayurveda along with Calvin Klein Jeans; the third boasts of Manish Arora and Gucci.

High-end malls like Emporio are frequented by Delhi's old elite, but perhaps their more important function is as a rite of passage for the new elite of the city and its hinterland. To be seen, and to make purchases, at Emporio is to take part in cosmopolitan, high-end global consumption. It is a way to get over one's provincialism and to stake a claim to power and status in the country's urban(e) core. The owner of an art gallery in Emporio noted that many of his customers come from smaller cities and towns, and have never bought art before, but see it as an opportunity to demonstrate their newfound worldliness.[41]

These dynamics also work at the lower end of the spectrum. Richer, more discerning shoppers want to be seen at the right malls, but for many who are just entering the middle classes, just going to any mall is enough to feel distinguished. For those living in the city, and even for those visiting it, a trip to a mall has become a kind of status-enhancing tourist activity.

Malls, then, create a new kind of society, breaking down old barriers and erecting new ones on the basis of a consumerism that is open to all, providing that one has the money. In some ways, state-promoted (though privately-run) malls have been more successful in crafting a new citizen than the more heavy-handed methods of the state-implemented (though private foundation-funded) Master Plan.

These social opportunities should not be dismissed as purely superficial and crassly commercial. Despite the failure of Gruen's socialist dreams, the mall still creates a new kind of social space and makes possible new kinds of social interaction, as the sociologist Sanjay Srivastava has convincingly argued. Srivastava has conducted extensive interviews with mall-goers in Delhi and Gurgaon, including the sons of farmers who have sold their land to the state or to real estate developers.[42] These young men, located on the cusp of the rural and the urban, see the malls as a way of taking part in the promises of a new India.

However, this social space is always a constrained one, because the ultimate goal of any mall is to make money. In Delhi, this means they exclude the majority of the population, who can't afford mall purchases. But the focus on profit also impacts those who do venture into the malls. Mall designers are faced with a dilemma: on the one hand, they must create the kind of atmosphere that draws people in and encourages them to socialize, but on the other hand, they must motivate people to buy products and not linger endlessly. One way of addressing this problem is named after the long-suffering Gruen, although he disavowed such techniques: the "Gruen effect" of giving malls an intentionally confusing lay-out, so that consumers get disoriented, lose their original purpose, and are more inclined to make random purchases.

There is something seductive about malls, and this seduction cannot be completely disregarded. It is based on promises that are not entirely false, just incomplete. Malls do indeed offer new opportunities, and they do create new, unexpected social worlds, but the liberation they promise is severely limited since it cannot be independent of the act of buying and the show of consuming.

Greenwashing

More easily dismissed are other promises made by the Vasant Kunj malls, particularly those of eco-friendliness. It's one thing to build a biodiversity park in a belated, PR-influenced recognition of the Vasant Kunj Ridge's ecological value; it's quite another to argue that the malls are actually good for the environment. This is precisely the claim made by the Environmental Impact Assessments (EIAs) that the project proponents hastily submitted to the Delhi Pollution Control Committee (DPCC) after the courts started breathing down their necks.

The EIAs are fascinating documents. Despite the air of secrecy surrounding the development of the malls, the EIAs are now in the public domain, albeit buried in the depths of the DPCC website.[43] The most immediately striking feature of the EIAs is the considerable space they devote to issues that seem outside the scope of environmental impact assessment. These sections are concentrated at the beginning of the EIAs and, though their content is extraneous, their purpose is clear: to prime the reader to be favorably inclined toward the project proponents.

Some of these introductory remarks are incredibly broad, presenting the familiar theme of Delhi as an important capital. As the EIA for the ONGC office complex intones: "Delhi, the capital of India, has always occupied a strategic position in the country's history." What is more, "its many-layered existence is tantalizing and can entice the curious traveler into a fascinating journey of discovery."

Some of the remarks are rather more specific, as with the preface to the EIAs prepared for Beverly Park Maintenance Services and Regency Park Property Management Services, representing the two DLF malls. Both EIAs begin with identical texts about the glory and prestige of DLF, as well as its downright patriotic roots:

> The saga of DLF... began in 1947 when India witnessed a huge influx of people resulting in doubling of the population of Delhi overnight. The Government at that time was hit by the need to provide shelter for this burgeoning population. DLF took up the challenge.

DLF was actually founded in 1946 and from the start engaged in intense speculation that drove up land prices, but this of course is not as endearing as a 1947 birth as a charitable organization.

But the most telling introductory remarks come from the Ambience EIA. Perhaps aware of the controversy around the mall, the EIA takes pains to clarify why the project is important. "The need for the proposed project arises to meet the commercial space requirements of the city and to stop the menace of unauthorized commercialization of the residential areas with the retail revolution that has spread geographically in India." This is a remarkable statement, arguing that it's acceptable to build on the Ridge, since this will stop the rise of other illegal commercial establishments. If a real estate project is big enough, the EIA seems to say, then it cannot possibly be illegal or illegitimate.

The EIA continues, "Fuelling this growth are India's sprawling shopping malls, which are increasingly challenging high street stores, corner shops and village market alike." Here, "sprawl" seems to be used in a positive way. There is no evidence given to suggest that malls have objective advantages over smaller shops or markets, or that they even contribute more to the economy. In fact, malls' relatively small role in the retail sector is briefly noted, but quickly followed by the bullish prediction that the "organized" retail sector would grow by 97 percent in the next five years, yet another hope that was dashed by the 2008 crash.

Having established the importance of Delhi, of DLF, and of malls, the EIAs then get down to the actual matter at hand: environmental assessments. Given their preludes, it is hardly surprising that the EIAs give the projects a clean chit. For example, the Ambience EIA notes that the area is not a Reserved Forest, which is technically true. It then goes on to note that it is not on the Ridge, which, at least geologically speaking, is patently false. It also gives the more ambiguous claim that the area is not "ecologically sensitive", but is, in fact, "ecologically insignificant". This is because no endangered species were found on the land, and because there was relatively sparse vegetation.

But even the environmentalists fighting the malls recognized that the Vasant Kunj landscape had been disfigured by decades of mining. It may not be ecologically sensitive, in that a surprising amount of dense green foliage sprang up after mining stopped. But the issue at hand was precisely how this land, now in the process of environmental regeneration, should be used: as a budding forest and protected water recharge area, or as a sprawling commercial complex?

The EIAs claim that the malls and offices would not affect the local groundwater tables, since they would be getting water piped in by the Delhi Jal Board, and, if that failed, they would get water supplied by tankers. But, even by the EIAs' conservative estimates, the amount of water needed for the complex was staggering: 1,616,000 liters per day. The EIAs describe plans to collect rainwater and recycle wastewater, but these are only mitigation measures that could, at best, meet a fraction of the malls' rapacious demands. Even if the water was not directly taken from underneath the Vasant Kunj Ridge, it would still be draining the city's already threatened water supply. The EIAs cover this up with bland platitudes: "Water conservation is an important part of sustainable living" and "The project proponents exhibit a general concern for water conservation and desire to operate in sustainable ways that would minimize any environmental impact."

The EIAs then go on to argue that the complex would, in fact, have a positive impact. How? The term "green belt" is used liberally. One EIA notes that the DDA would create a 50-meter-wide green belt separating the two sides of the main road parallel to the mall. The DDA has, in fact, built a wide divider on the road; about half of it has been greened, but largely with *vilayati kikar*. The other half is full of construction equipment, as work crews dig through the rocky ground to lay electrical cables.

The EIAs also emphasize that a "green belt" will be built around the malls. In practice, this has amounted to heavily manicured lawns, with carefully trimmed hedges, rows of flowers, and the occasional tree. This is just the kind of high-maintenance park that most ecologists disdain, as it requires considerable upkeep, including constant watering. It is, much like the forestation around Raisina Hill, window dressing, an aesthetic accoutrement to enhance the feeling of grandeur.

Perhaps the most unusual claim is the DLF Promenade's promise to build a green belt "in the mall". The interior of the mall is devoid of greenery; perhaps this was supposed to refer to the other DLF mall, Emporio, which features four massive palm trees, planted in huge pots, sitting inside the mall's posh atrium. It is less a green belt than a sparse quadrangle.

The EIAs emphasize that the malls will not just be good for the physical environment, they will improve the "socio-economic environment" of the area by driving up real estate values (which will be good for homeowners but not for renters), creating jobs (however precarious and low-paying they may be) and improving "the aesthetics and visual appeal of the region... by providing a cleaner and environment friendly office and commercial area". As the ONCG report sums up, "the overall impact on Socio-Economic Environment is positive and permanent in nature."

Those fighting the malls had quite a different assessment. One activist pointed out that the electricity used by the complex would be enough to power the homes of 80,000 middle-class residents. Others had more personal objections. Those whose homes had been demolished to make way for the Biodiversity Park pointed out the hypocrisy of clearing one small settlement on environmental grounds, only to construct a gargantuan complex that would have a far bigger environmental impact.[44]

These arguments were aired during contentious public hearings in 2006, soon after the EIAs were released. They were not enough to sway the courts. Even more disturbing, though, is that the EIAs are describing the best-case scenario, a rosy future world in which real estate developers and big corporations known for corner-cutting would follow environmental regulations with the utmost care.

But there is already evidence that this is not happening. In 2013, the National Green Tribunal found that the Ambience mall on the Vasant Kunj Ridge had not followed the conditions laid out in its EIA. Specifically, it had converted the basement, which was to be used solely for parking, into a commercial zone. It had altered other areas as well. The tribunal found that the mall had extended its commercial area by almost 60 percent, which would inevitably lead to more electricity use, more water consumption, more waste and so on. In response, Ambience argued that they would lose money if they cut the mall back down to its mandated size. The tribunal retorted, "Financial burden cannot be the consideration for compromising the environmental and public health interests."[45]

This was a welcome ruling, but for the Vasant Kunj malls, it was too little, too late. The Supreme Court ruling of 2006 had made it quite clear that, on the whole, the state machinery considers the balance sheets of private companies to be more important than environmental concerns. It is this logic that has fueled the growth of malls, on the Ridge and elsewhere.

Waste

Delhi's seemingly endless growth has created different kinds of surpluses: economic surplus for the nouveau riche; a surplus population to feed the growing service sector and construction industry; and surplus leisure time and surplus avenues of consumption for a privileged section of the population. This has been accompanied by corresponding shortages: a shortfall of reliable, decent-paying jobs for the majority of the population; a shortfall of housing for all (and especially for the poor); and a shortage of time for those desperately seeking work while trying to support a household.

To this list we can add another pair: a surplus of waste, and a shortage of places to put this waste. One environmental activist calculated that the Vasant Kunj commercial complex would, as a whole, discharge over a million liters of sewage each day, along with 13,500 kilograms of solid waste. And this is just one of Delhi's many malls.

Contemporary Delhi and its Ridge-top malls may be an extreme example, but cities, with their concentrated populations, have always produced a corresponding concentration of waste. Zooming out historically and geographically can help us better situate the current waste dilemmas facing the Ridge.

Historically, urbanization has led to environmental problems both in the city and in the countryside, especially in the case of what's politely called "human waste" or excreta. Shit (put more bluntly) is good fertilizer; it is not actually waste if, after proper treatment, it returns to the soil and nourishes the land. And this is indeed what happened for most of human history.

As cities grew, especially in the aftermath of the Industrial Revolution, the growing rift between town and country hurt both locales. In the towns, there was too much shit; in the country, not enough. As agriculture intensified, more was demanded of the soil, but its nutrients were not being replenished. This led to bizarre episodes like the guano boom of the mid-nineteenth century, when countries like Peru exported massive amounts of guano (bird shit) to European countries whose soils had become dangerously depleted. The European soil crisis was eventually resolved with the invention of artificial fertilizer, but this innovative measure did nothing to address the root cause of the problem: the increasingly stark division between city and countryside.

In cities, waste kept piling up. In both the East and the West, shit (euphemistically called 'night soil') was generally removed from the city manually, by sweepers, who loaded the shit onto carts and then disposed of it outside of town. The link with agriculture was not totally severed. The shit was generally deposited in large pits, composted, and sold to farmers. This was the dominant system used in Paris and

London, for instance, along with Indian cities. Everywhere, this was considered dirty work, but it was especially stigmatized in India, as it was tied to age-old hierarchies and caste-based exploitation; this was considered the polluting work of the castes deemed untouchable by their high-caste oppressors.

In seventeenth-century Shahjahanabad, the manual system of night soil removal was supplemented by a basic network of subsoil sewers, which was occasionally flushed out with water from a nearby canal. By the time the British took control of Delhi, though, this sewer system was already in disrepair, another victim of the Mughals' long decline.

British rule in India was marked by an obsession with sanitation. The British set up their own kind of caste system, increasingly paranoid about any kind of contact with a "native" population seen as impure and disease-ridden. In Delhi, this paranoia perhaps reached its zenith (or, more accurately, its nadir) when the Army's General Quarter Master recommended that Indian shit and British shit be disposed of in separate trenches.[46]

This did not happen, although other kinds of segregation certainly did, first with the creation of British Civil Lines as distinct from Shahjahanabad, and then, more dramatically, with British New Delhi prioritized over native Old Delhi. Both sides of the segregated city still produced shit in increasing amounts as the city grew, and this needed to be disposed of. As early as the 1870s, the sites used for dumping and composting shit, which were located in the immediate outskirts of the city, had reached their capacity. So the city turned to a space that had long been labeled as 'wasteland' by the British: the Ridge. What better place to store human waste than a place that was itself perceived to be waste? Ridge land also had the advantage of being sparsely populated and agriculturally unproductive; it thus became a favorite dumping ground for the British and eventually for the post-Independence managers of Delhi's waste.

From the 1870s onwards, cartloads of shit were carried to increasingly far-flung Ridge locations, starting just outside the walled city and eventually reaching the outskirts of Qutb Minar, many kilometers away. This caused problems later, as the British tried to expand the city's infrastructure. In 1903, for instance, the British government in Delhi was considering buying a tract of land just south of Paharganj (literally, "hilly neighborhood") to build housing for government workers tasked with managing the region's growing railway infrastructure. In a hand-written note on the request for acquiring the land, an official wrote, "This is the very piece of high ground where sewage was buried some years earlier." Another added, "Inform the railway that filth was buried all over this plot about three years ago."[47] The city was having a hard time escaping its shit.

As sanitation conditions worsened, the British began developing a piped water system, which would be used for both the delivery of drinking water and the flushing out of shit; in short, the modern sewage system that is now a unremarkable staple of our daily lives, but was, at that time, still novel even in most of Europe. In Delhi, as in other colonial cities, the benefits of this system were distributed extremely unequally, as the system catered to British residents and, later, a handful of elite Indians. This inequality was exacerbated by the construction of New Delhi and the concentration of urban infrastructure, including sewerage, in the new city, at the expense of the old.

The vast majority of Indians in Delhi, then, did not enjoy the new "water closets" built by the British. They largely relied on public latrines, but even these were built in insufficient numbers, especially when the British started channeling their energies into New Delhi. This meant, for many Delhi residents, especially those living in the city's outskirts, that the only option was shitting outside, "open defecation", that demon that the Indian government is still trying to exorcise.

British officials, with their sanitation obsession, found this quite alarming. For the most part, their solution was not to build more latrines, far less to extend the sewer system. It was to police the "native" parts of the city with ever increasing vigilance, as the colonial archives amply demonstrate.

In 1872, for instance, British officials in Delhi started to complain about a site on the Northern Ridge near Hindu Rao's house, where locals were supposedly encroaching, and were using a local drain as their toilet. In the quaint, euphemistic British English of the times, officials bemoaned the "unsanitary" nature of the drain, since locals would "ease themselves all around it." The main agencies involved in this case were the army and the city's Municipal Committee. The land in question belonged to the army (a legacy of the 1857 Uprising), and civil authorities wanted it transferred to them so that they could manage the draining of the area and improve its sanitation.[48]

The exchanges about this issue are filled with the usual bureaucratic tussles over jurisdiction, as well as the usual politely worded barbs aimed at rival officials. Although sanitation was the main issue, time and money were never far from the minds of the officials. One wrote drolly to his counterpart: "I cannot agree with you that, 'no great delay seems to have occurred in this case which calls for explanation,' unless it be by way of contrast with other cases." Later, the same official, in high dudgeon, decried the "continuance of the nuisances in drainage and conservancy..., of the encroachments and their non-removal, and other glaring evils... They are all remediable if only the matter be taken up with energy, perseverance, unity of purpose and mutual help." He was to be sorely disappointed.

The army was unwilling to part with the land, but they still wanted the Municipal Committee to improve the infrastructure ("for the public good," military officials hastened to add). Letters flew back and forth, and money emerged as the crux: if it couldn't own the land, the Municipal Committee felt it should at least be able to collect taxes on it. The military refused.

Officials could not even agree on the root cause of the problem, although they all agreed that "natives" were shitting on army land. One official confidently stated: "Since the removal of a latrine nearby... the people residing in that quarter have no place where they can resort to and it is therefore not to be wondered that they should ease themselves wherever they could find a convenient spot." Another, though, countered that the site in question "is too far to be affected by the removal of the latrine."

As officials debated the precise cause of the problem, they also disagreed on an appropriate solution. Some thought that the "encroaching" huts should be demolished; others demurred. On one issue, though, there seemed to be consensus: the land should be leveled. Besides the contours of the Ridge, the landscape was pockmarked by holes from quarrying, which was prevalent even at this early date. Several officials used aesthetic arguments to decry quarrying, saying that it ruined the picturesque summit of the sacred Northern Ridge. But the more serious problem was sanitary; the quarried pits were being used as toilets.

Leveling the land would remove these impromptu latrines. It would also have other benefits. As one official noted, there had been less "nuisances" in the area,

> but this will last only so long as strict watch is kept over the locality. The surface of the ground is so uneven that it is difficult to keep up this watch so that in course of time it will relax.... But if the space be leveled it will be impossible for persons to defile the place without being seen from a considerable distance.

It was the same basic logic that Balban used when cutting down all the trees around Hauz-i-Shamshi: create a clear line of sight, and thus improve the state's surveillance capabilities.

Here, however, the archival record stops. We don't know if the land was ever transferred to the Municipal Committee, or if the huts were destroyed. The area in question is still quite hilly, so it's unlikely that the British took any drastic land-leveling measures. Indeed, given the persistence of in-fighting between officials and the abrupt end of the official correspondences, it seems probable that little action was taken.

We do know, though, that the colonial logic of sanitation and surveillance continues to this very day. Perhaps the most gruesome example of this was a violent incident in 1995, when residents from a middle-class neighborhood of Ashok Vihar, along with two police-men, chased down an eighteen-year-old man, whom they suspected of using a neighborhood park as a toilet, and beat him to death. When residents of the nearby working-class neighborhood pro-tested, the police again resorted to violence, shooting and killing four people.[49]

Such intense violence says much about the callousness displayed by many of Delhi's wealthier residences. It also suggests that Delhi has not come any closer to solving its sanitation problems. On the contrary, with the city's ever-growing population and the DDA's persistent failure in providing adequate housing, the problem of human waste has only become more severe. The Ridge continues to be a favorite location for people who need a place to shit.

In many Ridge areas, in the early morning, it is common to see men and women from nearby settlements streaming into the Ridge, water bottles in hand, searching for a spot to relieve themselves. As more of the Ridge has been set aside as parks and forests, it has become even more attractive as a makeshift toilet, since it offers shade and relative isolation. Sometimes this practice has even been institutionalized. At a shrine in the Mehrauli area of the Ridge, water mugs are placed at the edge of forest, with a big sign in Hindi saying, "TOILETS", with an arrow pointing to the left for ladies and right for gents.

Shit is not the only waste dumped on the Ridge, even if it's the one that has induced the most medical and moral panics. In a city where the real estate market has played such a dominant role, construction waste has been produced in ever-increasing amounts. In 2016, Delhi produced roughly 3000 tons of construction waste per day. And a significant part of this waste gets dumped onto the Ridge, simply because of its convenience as an isolated, non-populated space. The National Green Tribunal, in response to PILs, has tried to address this issue, but its rulings usually have little effect on the ground.

The problem is not just limited to private construction companies. The state itself has been a major culprit in improper construction waste disposal. In 2004, the DDA was caught dumping 5000 tons of construction waste on the Ridge. In 2009, the Central Public Works Department was also caught in the act, emptying rubble into small ponds on the Southern Ridge. These sites have eventually, grudgingly, been cleaned up, but the agencies involved have received no punishment.[50]

Delhi's waste problem is, of course, not just confined to its human waste and its construction waste. It produces massive amounts of waste, period. If it is not illicitly dumped on the Ridge, then it likely makes its way to one of Delhi's massive landfills. But just like the shit pits in 1870s Shahjahanabad, these landfills are overflow

Delhi's dumps have rewritten the contours of the city. Earlier, Delhi's topography was largely defined by its geology: the Ridge, of course, along with two byproducts of the distant Himalayas, namely the Yamuna River and its alluvial soil. This geological base has, over the years, been significantly modified, with parts of the Ridge leveled to make way for new neighborhoods, and other parts quarried to create pockmarked landscapes.

Meanwhile, landfills are creating new hills and reconfiguring the landscape. From a distance, they look like looming mountains, far taller than the crests of the Ridge. Closer up, their solidity melts away, as pieces of garbage shift, merge together, disintegrate, topple. The three functioning landfills of Delhi are packed to the brim, and they still receive roughly 9,000 metric tons of garbage a day. At the core of these dumps, the waste is so densely crushed together that it is breaking down and releasing high levels of methane. Fires break out with alarming frequency.

The Municipal Corporation of Delhi (MCD) is desperately searching for new landfill sites in the city, but no one wants a new dump in their backyard. The MCD has repeatedly suggested turning part of the Bhatti Mines into a landfill. There is something poetic about this vision. The craters of Bhatti were created to feed the construction boom in Delhi. Now that they're closed, the holes can be filled up with the waste of those living in the city's countless new buildings.

Environmentalists, the Delhi High Court, and even the Delhi Government have opposed this plan, even as the World Bank has promoted it. For the time being, the MCD has shelved the plan, perhaps waiting for the controversy to die down. Meanwhile, in 2014, the DDA cleared seven new sites for creating landfills. None of these sites are in Bhatti; two, though, are further down the road, in another part of the Southern Ridge near Maidangarhi, one of the many villages whose common lands were taken to create a Reserved Forest in 1996.[51]

This is one more underside of Delhi's explosive economic growth. Piles of shit; piles of construction waste; piles of undifferentiated solid waste creating a new map of the city. The Ridge has fed Delhi's growth: its thorny scrub forest has fed the city's livestock; its quartzite has fed its magnificent palaces; its Badarpur sand has fed its roads and its concrete buildings; its rising land values have fed the rapacious real estate industry. It is only fitting that the waste from all these processes finds its way back to the Ridge.

5 SPIRITS
Transcendence, Sacred and Secular

Purity/Pollution

In the morning, the Northern Ridge's Kamla Nehru Park is a bustling place. With its neatly manicured hedges, its spotless paths and its proximity to Delhi University, it draws many walkers, joggers, and yoga enthusiasts. There is even an open-air gym for fitness buffs. Next to the gym is a small Hanuman Mandir, appropriate given the proliferation of monkeys in the area. Like most of Delhi's mandirs, this one claims to be *pracheen* (ancient), a view contested by the caretakers of the gym, who are in a longstanding feud with the temple pujaris over the boundaries of their respective institutions.

I visited this temple one morning along with a friend, and we tried to interview the head pujari. He was sitting serenely, in a spotless white kurta and dhoti, a mala around his neck. When I mentioned that I was interested in environmental issues, he became agitated. Before I could finish describing my research, he exploded, "You want to talk about pollution?! Then talk to all these boys and girls who come here together. They're the ones who are really polluting the place! Ask the Prime Minister why they're spreading pollution!"

The pujari was referring to the young men and women, many of them college students, who come to Kamla Nehru Park for romance. If mornings are for exercise, then the languid afternoons are for love. Couples sit on the park's benches or on the grass between the trees, holding hands, whispering to each other, sneaking kisses. If the couples want more privacy and intimacy, they slip a hundred rupees to the security guard patrolling the area, and he lets them retreat to a more secluded, densely forested part of the Ridge, where he leaves them in peace.

The security guard, as long as he gets his payment, seems to have few moral scruples about the arrangement. The pujari, on the other hand, sees this as a kind of spiritual contamination, sullying the positive aura of his temple. And he can draw on a long tradition of demonizing the forest, and all the impurities and dangers it contains, a tradition that still reverberates in the present day, on the Delhi Ridge and beyond.

For instance, the ancient legal text *Manusmriti* advised kings to establish their kingdoms in dry, open areas, with sparse vegetation—not coincidentally, the kind of savanna ecosystem that was ideal for pastoralism, and that was often actively created through the use of fire. The civilized, largely pastoral area was lauded as ritually pure. It was contrasted with the *anupa*, or wet, wild forest, the abode of barbarians and fearsome beasts, a horribly impure zone.[1]

The purity/pollution distinction did not just work to separate the barbarous outside from the civilized inside. It also created hierarchies and social controls within the bounds of civilization. Caste hierarchies, for instance, were (and are) justified with the assertion that the "higher" castes are the ritually pure ones, the "lower" castes ritually impure. Women, too, especially menstruating women, are seen as impure, and thus inferior to men. The oppressive logic of these hierarchies is distilled in the *Manusmriti,* a text that presents a Brahminic vision of an ideal society and consistently associates marginalized groups with untamed, impure nature.

In such visions, "nature" was shunned in part because its power was feared. Gail Omvedt notes that early caste-based societies "had... a notion of sacred powers in nature which were potentially dangerous, and a conceptual linkage of these with certain occupations and activities and with women." Over time, "the 'dangerous' became the 'polluting', and eventually 'impure' and 'low'."[2]

However, the process of taming "nature"—whether by pushing it to the fringes of civilization, as with "barbarous" populations, or by punishing and demeaning it in the heart of civilization—has always been a difficult, incomplete project. This is both because of the resistance and rebellion of oppressed peoples, and because of the ultimate impossibility of definitively separating humans from nature.

Further, while Brahminical traditions in India have sought to create gender and caste hierarchies through spuriously naturalized notions of "purity" and "pollution", there have been a proliferation of other spiritual traditions in India, ones far less concerned with social distinctions and far more embracing of nature. The tradition of wandering sadhus, of yogis and yoginis in forests and caves, is a potent reminder of the sacred power of the wilderness. These wanderers have typically had no concern for caste and social convention, preferring the freedom of a nomadic lifestyle.

The tension between the civilized and the wild, the orthodox and the heterodox, the rulers and the rebels, has long been felt in the Delhi region. The coexistence of uptight pujaris and furtive lovers in the present-day Ridge is evidence of this. But the fight is not just between secular students and the sacred guardians of morality. It is a battle to define what constitutes the spiritual itself, with the moralizers facing off against the mystics. Not that this battle has fixed formations. With the vicissitudes of time and the compulsions of more worldly politics, mystics have often become moralizers, and vice versa; and spiritual movements traversing that Ridge have often struggled to contain contradictory impulses.

The history of transcendence on the Ridge is thus a complex, bewildering one, which stretches from the mythical past to contemporary times. In an attempt to tackle this complexity, this chapter is divided into three parts, though the borders between them are inevitably blurry. The first looks at transcendence within the realm of religion, of the mystics of various stripes who have played a prominent role in the Ridge's history. The second looks at more 'secular' attempts at transcendence, as the lovers of the Ridge make a return, along with drinkers, revelers, criminals and down-on-their-luck royalty. The final section recognizes that the division of the previous two parts of the chapter is ultimately an artificial one: sacred and secular spirits co-mingle, as they are brought down to earth and into interaction with the subjects of the previous chapters.

PART I: SACRED SPIRITS

Rebels on the Ridge

One figure that pops up with surprising frequency on the Ridge is that of the Nath yogi. These holy men are often referred to as *kanphata* (split-eared) yogis, due to the round wooden rings that pierce a large hole in their ears; the yogis themselves prefer the term "Siddhas" or "perfected ones". The historian Narayani Gupta mentions the presence of Nath yogis on the northern end of the Ridge, both in pre-British times and in the present day, "with their trademark black blankets" and "their camping-places and shrines".[3] In recent years, Nath yogis have had an expanding presence on the southwestern portion of the Ridge, specifically in the Reserved Forest area now called Sanjay Van, an area we will visit several times in this chapter. And finally, Nath yogis have a long historical association with the Kalkaji mandir on the slopes of the southeastern Ridge, in a portion of hills that have not earned government protection as forest. Even today, Nath yogis perform Tantric aartis for Kali, the deity worshiped at Kalkaji.

The confluence of Nath yogis, Tantra and Kali on the Ridge is suggestive. Though representations of Kali vary greatly over time and space, the goddess is undoubtedly associated with fierceness, unruliness, indomitable wildness. These resonances are highlighted through Kali's pride of place in Tantric traditions. Tantra's relationship to more orthodox traditions roughly parallels the relationship between the forest and more civilized spaces. That is, Tantra offers the possibility of escaping the stuffy confines of societal norms, but in doing so it is simultaneously liberating and dangerous. Tantriks often come from humble backgrounds, and Tantric traditions generally have little regard for the caste boundaries that orthodox traditions have so painstakingly erected. Tantra also offers the promise of spiritual transcendence in this very life, and not in some distant future. But yogis and tantriks are associated with all kinds of taboos: sex, drugs, violence, black magic and much more.

The mystery around Tantra is compounded by the fact that most Tantric traditions are intensively secretive, in part because of the suspicion their activities arouse. It is also difficult to make any generalizations, since Tantra is a diverse tradition spanning many centuries, and since many once-transgressive traditions have been cleaned up and incorporated into more orthodox, hierarchical practices, thus gaining societal acceptance but losing their radical edge, as is the case, we will see, with some sections of the Nath yogis, despite their Tantric roots.

Whatever the present-day compromises they have made, Nath yogis have long practiced *hatha yoga*, or literally, "the yoga of violent exertion". Through the solitary practice of *hatha yoga* techniques, Nath Siddhas are able to achieve superhuman feats, or so they claim, traveling through time and flying from mountaintop to mountaintop. Their ultimate goal is to achieve bodily immortality. These legends still hold today, as siddhas and yogis dot the landscape of both rural and urban India, including Delhi's own Ridge.[4]

Tracing the Nath Siddhas through Delhi's history, alongside Sufis and goddesses and other mystics who have frequented the Ridge, reveals the tenacity of the wild and the transcendent in the city, as well as the difficulty of maintaining airtight categories and binaries: nature vs. culture, spiritual vs. material, wild vs. civilized. Siddhas, yogis, Sufis and goddesses have knowingly blurred the lines between these categories, drawing on the power of wild nature to intervene in "civilized" affairs, often in historically decisive ways.

Mystics and Goddesses of the Arid Zone

The life of Gorakhnath, the original Nath Siddha, is shrouded in mystery. His historical existence is obscured by competing, overlapping legends and mythologies; he is often portrayed as immortal, disappearing and reappearing at different points in history, but it seems his lineage emerged sometime in the eleventh or twelfth century.

This suggests that the Nath tradition was born during the turbulent days of Arid Zone dynamism, a time that brought Delhi and its Ridge to political prominence (as we saw in Chapter 2). The Nath yogis were just one of many heterodox sects that gained prominence in this time, and that made use of the expanding networks of trade, piety and plunder throughout the semi-arid borderlands of India. In these days, yogis mixed freely with Sufi mystics, exchanging techniques and building a new set of popular devotional practices.

But the yogis did not restrict their activity to the religious realm. During this time, the line between mystic, mercenary and merchant was often quite indistinct.[5] This was aided by the secrecy of many Tantric sects, which made it difficult to distinguish between a genuine and a spurious yogi. It was thus easy for merchants or warriors to don a yogic guise in order to ease their travels and avoid detection when necessary. The reverse was also true, as yogis searched for ways to make money and put their physical and spiritual training to use.[6]

Even when they did not directly hold economic or political power, Nath Siddhas were seen, in medieval India at least, as the quintessential king-makers. This was, again, due to their grounding in the Tantric crafts. They had, it was believed, achieved a mastery of the material world, which gave their blessings and curses great weight for those seeking to build an empire.[7] Their closest analogue in the medieval West is probably the wizard. And just as Merlin propped up King Arthur, so too the Nath Siddhas supported Arid Zone kings.

The mythical links connecting Nath Siddhas to Delhi start with the pivotal battle between Prithviraj Chauhan and Mohammad Ghori. As we've seen in Chapters 2 and 3, Ghori's victory over Prithviraj has been given outsize importance by later Hindu nationalist commentators, who imagined the early fortresses of the Ridge as key (if ultimately ineffective) barricades against foreign invaders. Circling back to this battle, we find yet more suggestions that this was not a simple struggle between the righteous Hindu Indian king and the diabolical Muslim outsider. Rather, alternative versions of the story, found in folklore and oral narratives, reserve pride of place for heterodox tantriks flexing their supernatural muscles.

Specifically, folktales tell of two Nath Siddhas, Guga and Ratan Baba, both of whom had ties to mystical Islamic practices, and both of whom, in mysterious, idiosyncratic ways, aided Mohammad Ghori in his march to Delhi.[8] Ratan Baba demonstrated his yogic greatness by producing a magical water bowl that quenched the thirst of Ghori's entire army, no mean feat in the perilous Arid Zone. He then sent Ghori off with his blessings. Ghori then visited Guga's tomb, which was poorly maintained. Troubled by this, Ghori vowed that, if he was victorious, he would return to the tomb and restore it to good condition. With the backing of Guga and Ratan, the hybrid yogi/sufis, Ghori went on to defeat Prithviraj and take control of his Ridge-top fortress, as well as his other territories. He then kept his promise to Guga, repairing and renovating his tomb.

If these folktales suggest that Mohammad Ghori may not be the outright villain of 1192, other stories, from regional martial epics, indicate that Prithviraj Chauhan may not be universally regarded as a hero, even by other Hindus. For instance, Prithviraj is shown in a far-from-flattering light in an epic known as the *Alha*, which focuses on the intra-Rajput disputes that weakened North Indian kingdoms before the arrival of Mohammad Ghori. The *Alha* is a fascinating tale that undercuts the legend of Prithviraj while shining light on new characters: lower-caste heroes, strong women with links to the divine, and holy men who are not what they seem. The *Alha* also redraws the lines of military conflict in this era: the battles of the epic are not portrayed as Hindu versus Muslim, but rather as a fight between minor kingdoms, often associated with low-status populations, versus growing empires, confident in their elite superiority.[9] And in a strange, circuitous way, the story of the *Alha* eventually brings us back to the question of transcendence on the Ridge.

The heroes of the *Alha* come from the small city of Mahoba in present-day Madhya Pradesh, then the capital of the Chandel Rajputs. Mahoba was the center of a small kingdom that got swept up in larger inter-imperial battles. For the protagonists of the *Alha*, Prithviraj is the big bad enemy with imperial ambitions. He threatens to swallow up their little principality and has little regard for their claims to sovereignty. For minor kings like the Chandels, Prithviraj is as bad as Mohammad Ghori—both are imperial schemers with expanding ambitions, and their differing religious backgrounds hardly matter. This point is underscored by the Chandel's alliance with the Muslim statesman Mira Talhan, who was also trying to fend off the imperial ambitions of Prithviraj.[10]

The titular hero of *Alha* is not the Chandel king but rather one of his trusted lieutenants, who hails from the Banaphar clan. The Banaphars also identify themselves as Rajputs. Throughout the epic, though, they have various caste slurs hurled at them by higher-status Rajputs who claim that the Banaphar line is contaminated with the blood of Ahirs, a nomadic pastoral community. These slurs underline the hardening of caste boundaries, as royal Rajput groups, who themselves had emerged from pastoral or tribal communities generations earlier, cordoned themselves off from nomads at the periphery of their empires. The Banaphars are an embodiment of the wandering warriors of this period, who sought (though often failed) to raise their status by taking on a mix of military and mystic guises. At one point in the epic, four Banaphar brothers, along with Mira Talhan, disguise themselves as warrior-yogi-musicians so they can take revenge against a clan that murdered several Banaphar elders.

But this is just a side-story in the epic. The main focus is the fight against Prithviraj Chauhan. The extent versions of the *Alha,* which have been told and retold over many centuries, place Prithviraj in Delhi, even though the Chauhan king did not actually rule from here. But Delhi's centrality as an imperial capital (which only took shape during the reign of Iltutmish) has rebounded through time, and in the epic, it's the natural home for Prithviraj. The Delhi-based king earns the resentment of the *Alha's* heroes when he refuses to go through with his daughter's proposed marriage to the Chandel prince Brahma.

The daughter, Bela, is no mere princess. She is, in many tellings, a reincarnation of the *Mahabharata's* Draupadi, who is herself revered in many parts of India as a reincarnation of Kali, who is known to drink blood and even to eat corpses. And even though Bela is Prithviraj's daughter, she is, in the *Alha,* symbolically and thematically linked to his enemy the Banaphars, as the goddess who strives (in the end, unsuccessfully) to defend their land.

Bela's role in the *Alha* mirrors the role of Draupadi in folk retellings of the *Mahabharata.* Both women are, like Helen of Troy, the immediate cause of a war. But their position is much loftier than that of mere instigators. As incarnations of Kali, both Draupadi and Bela oversee a divinely ordained destruction, sweeping away petty, squabbling kings and clearing the ground for a new era. In the case of the *Mahabharata,* this new age is the dreaded Kali Yug; for the *Alha,* it is the age of Central Asian incursions and Muslim influence.

In the *Alha,* Bela's main enemy is her father. Prithviraj keeps postponing Bela's marriage to Brahma, eventually luring the prince to come to Delhi and try to take Bela by force. One of Prithviraj's advisers, the deceitful brahmin Chaunra, dresses up like a woman and pretends to be Bela. Upon meeting Brahma, Chaunra stabs the unsuspecting prince with a poison-tipped dagger. Fatally wounded,

Brahma retreats, but only makes it halfway to Mahoba. He clings to life, though, for the next few days.

Learning of this treachery, two of the Banaphar brothers travel to Delhi, posing as freelance soldiers who want to work for Prithviraj. They infiltrate the palace and free Bela, who is desperate to see Brahma before he dies. Before she leaves Delhi, she curses it, damning her father and his meddling, and ridiculing his imperial ambitious. She says, prophetically, "There will be widows in every house in Delhi... A river of blood will flow in Delhi... a thunderbolt will fall on Delhi."[11]

In keeping with her Kali energy, she then presides over great bloodshed and gory battles. At Brahma's urging, she disguises herself as a male warrior and faces her brother in battle. Victorious, she beheads him and brings the head back to Brahma, so he can die in peace. She then sets herself on fire as the battle between Alha and Prithviraj rages around her. Both armies are decimated, and all the Chandel princes are killed. This spot is commemorated by a small shrine in a village near present-day Bhopal.

But Bela has lived many lives, and her final resting place is difficult to pin down. She appears in other stories as well, tales that have not been recorded as diligently as the *Alha*. These other legends spread more quietly, circulating by word of mouth, in stray textual references, and, very recently, emerging into the online world in the form of blogs and travel websites. It is one of these elusive narratives that brings us squarely back to the Delhi Ridge.

I quite literally stumbled upon this tale. Within Sanjay Van park, in the Mehrauli section of the Ridge, are ruins from the Tomar fort Lal Kot, crumbling remains of the oldest walled compound in Delhi. From the top of these old ramparts, there is a magnificent view of the nearby Qutb Minar, while the landscape below is densely packed with *vilayati kikar*. One day, following the path of the crumbling walls, I ventured deep into the forest and saw a clearing in the trees below. A rough rectangular courtyard had been created, bounded by a makeshift quartzite wall that provided an elegant framing for the two graves in the courtyard's center. The graves stood on a small platform, painted green, and were covered with colorful sheets.

The courtyard was empty, and I soon turned back, as my path petered out in a maze of shrub and thorn. But I returned several weeks later with a friend, and we were lucky to find three men near the courtyard, gathering wood from the surrounding forest. It was January, and the sun shined weakly through the winter haze. One of the men was preparing a small fire with *vilayati kikar* branches he had gathered. When the men saw us, they guided us down the steep crumbling path that led from the ramparts to the courtyard.

The caretaker of the graves was a migrant, in keeping with the proud Delhi tradition. He came from a small village in Bihar. He told us that, one night, he had an intense spiritual vision, in which he saw this very shrine and was told that he must come to Delhi and find it. So he did. He initially apprenticed under the elderly caretaker he found there. When the old man died, he took over the position. He now lives at the shrine year-round, though the chill of the winter and the fiery heat of the summer, sleeping next to the graves, giving *prasad* to visitors like us, guiding the slow trickle of pilgrims that come to see the shrines, and fending off the occasional snake or porcupine.

But who, we asked, is buried here? The caretaker responded that one shrine belonged to Haji Roz Baba, one of the first Sufi saints to come to Delhi, a great mystic and a venerated teacher. Next to him, said the caretaker, was the grave of Bela, Prithviraj's daughter, who had been initiated into Islam by Haji Roz Baba and had become a spiritual master in her own right.

Bela: not just Draupadi, not just Kali; now also a Sufi mystic. This Bela legend has little purchase in scholarly circles, although it was circulating in Mehrauli at least as far back as 1922, when Zafar Hasan compiled his list of Delhi's monuments.[12] Despite the legend's marginal status, it resonates with other Bela tales, including those told by tribal communities in Uttar Pradesh, which position Bela as the sister of Lakhan, a king from Kannauj, the former stronghold of the Pratihara Gujjars. In these tales, Lakhan goes to Delhi and converts to Islam, and he prospers due to the blessings he has received from Bela.[13] So Bela travels, changing shape and shifting religions, but always powerful in her blessings and her curses.

A part-Hindu, part-Muslim, part-human, part-divine prophet of tragic love and violent destruction, Bela deserves her location in the midst of Sanjay Van, in a wild part of the Ridge. It seems appropriate that the shrine is located outside the walls of Lal Kot. Like any true mystic, she could not be confined within the city's bounds. More broadly, Bela represents the mysticism and cultural mixing that spread through the medieval Arid Zone, and still lives on in the Delhi Ridge. This is especially true in the Ridge zones around Mehrauli, infused with centuries of history and dense with spirits.

Sanjay Van's Sufis

Bela and Haji Roz Baba are not the only Sufi mystics whose presence graces the Ridge, and particularly the part of the Ridge now referred to as Sanjay Van. Centuries earlier, this patch of land was located in the hinterland of the Ridge-top fortresses controlled by the Rajputs and then the early Slave Dynasty. As Delhi developed as an urban center, Sufis began to populate its hinterland. The sultan Iltutmish sought to contain the power of popular Sufis, including the famed Qutb Sahib, whose blessings and curses rivaled those of the yogis. Though Sufis were never tied to the kind of extreme taboo-breaking that characterized some Tantric sects, they still had an uneasy relationship with orthodoxy, especially as represented by the conservative clerics that often aligned with state power.

In many ways, the mystical Sufis had more in common with the Nath yogis than either group shared with orthodox Muslim clerics or orthodox Brahmin priests. Sufis were more open to sensuality and were known for their singing and trance-like dancing (the famous whirling dervishes). Like the yogis, they had little patience for social hierarchies.

This affinity was not just an ideological one. It had real grounding in meetings and exchanges between the two groups. The Sufis were attracted by some of the more inward-looking techniques of the Nath Siddhas, especially the breathing techniques highlighted in hatha yoga, and they soon incorporated this into Sufi practice.

The Sufis and the yogis also shared a love for the wilderness, and they were known for their meditative sojourns in far-off mountain caves and dense forests. In Mehrauli, local legend has it that Baba Farid, Qutb Sahib's disciple and Nizamuddin's teacher, liked to meditate in an isolated patch of jungle not far from B mb. He used to venture to this spot for 40 days of meditative s de. Near this spot is another shrine, this one belonging to Sheikh Shahbuddin. Little is known about Shahbuddin, but the present-day caretakers say he is Qutb Sahib's nephew.

The current guardian of the shrine, Akbar Shah, is considered an accomplished mystic in his own right, a "mazjoob" who is so absorbed in his spiritual trance that he can appear to outsiders as slightly mad. He never speaks more than a few words at a time, and he does not eat solid food, subsisting mostly on water and the occasional chai. According to the shrine's attendants, he too is a miracle worker. One day, the attendants at the dargah found that the shrine's well had become dry. In present-day Delhi, these kinds of water shortages are a chronic problem, as the population grows, elite consumption sky-rockets and the groundwater levels plummet. Akbar Shah meditated on this and said that water would reappear in the well the next day; and so it did.

Within a stone's throw of this well is a spot even more shrouded in mystery and mystic legend. Behind the shrine is a steep quartzite slope, and on the other side of this slope, a narrow, rough, rock-hewn staircase leads down to an isolated cave. The cave has a metal grate closing it off to the public, with the key kept on the premises of the nearby shrine. This cave is used for 40-day meditative trances, and is opened only for accomplished Sufis.

But its more esoteric purpose is to summon Khwaja Khazr, the legendary "Green Saint" of Islam. Khwaja Khazr, known in Arabic as al-Khadr, is at the center of a bewildering, ever-expanding set of myths and legends, which stretch back to the genesis of the world and spread across the globe. He appears as a guide for those who are lost, especially those stranded in the wilderness. He is immortal, it is said, and has provided solace to Moses and Alexander the Great and kept watch over the mythical Fountain of Youth. Spilling over into other faiths, he is believed to be an incarnation of Vishnu or the Zoroastrian deity Sraosha, or an alter-ego of John the Baptist. He has been tied to characters in the Epic of Gilgamesh and even in the Book of Genesis.

Wherever he appears, Khwaja Khazr is associated with verdure. When he kneels down and prays to the soil, shrubs, grasses and trees immediately spring up, truly miraculous for a sage of the Arid Zone. And he is known to haunt the Delhi Ridge, and specifically that corner of the Ridge now known as Sanjay Van. However, for decades now, Sufis have been unable to call up the Green Saint.[14] Perhaps the deathless sage is simply away on one of his international sojourns. Or perhaps he is the secret guiding force behind the greening of Sanjay Van, which has been selected by the government as a model city forest, and which is indeed quite verdant as a result of the plantation efforts of the DDA and various environmental NGOs.

Syncretism and Status in Mehrauli

Mehrauli's most popular Sufi, Qutb Sahib, is now no longer in the wilderness. Though his shrine is less than one kilometer south of Sanjay Van, it has been swallowed up by the urban settlements of Mehrauli. The shrine has been a popular devotional site since Qutb Sahib's death in 1235 CE, and it provides one of the few lines of continuity through Delhi's fragmented, fickle history. Qutb Sahib himself chose the location for his future tomb when he was passing through the jungle outside of the walls of the city. He was heading back from a Ramzan festival near Hauz-i-Shamsi, which at that time was still surrounded by dense vegetation and plagued by nomadic raiders and bandits. The fact that this now-bustling spot was once forbiddingly remote is indication of how tiny the original Delhi was.

Historical accounts of the shrine suggest that it has retained its wild side through the centuries, even in its now-urban surroundings. This is especially true on the occasion of Urs, the ritual celebration of a Sufi saint's death anniversary. In the Sufi tradition, this is generally seen not as death, but as union with the divine Beloved. Urs is not a time for mourning, then, but for celebration, more wedding anniversary than funeral remembrance.

Today, the Urs at Qutb Sahib's shrine draws a huge, heterogeneous crowd. The highlight of the Urs festivities is the performance of qawwalis, but there is something here for everyone. Kids bounce off each other in a huge inflatable playpen. Small stalls sell religious books and trinkets. Cooks prepare biryani and sweets of all kinds. Chaiwalas brew tea.

The massive crowd can be roughly divided into three com- ponents: itinerant Sufi mystics from various orders who travel from shrine to shrine; mainstays of the Sufi community, caretakers of important shrines, religious scholars, wealthy Muslim donors, and respected community elders; and, finally, ordinary devotees from Delhi and the surrounding region, who are drawn to Qutb Sahib's shrine largely because of the saint's healing presence. Many Sufi healers, drawing on the strength of the saint, gather here to tend to the physical and emotional wounds of those who visit. Like the yogis, the Sufi healers are considered masters of the material world, able to remedy chronic illnesses and relieve psychic torment.

Those searching for healing are generally (though not exclusively) poor, and the network of healing shrines and temples constitutes an alternative system of healthcare, which both competes with and complements the more mainstream medical system. Many attending Qutb Sahib's Urs have come to celebrate, but also to seek solace and find respite for their woes. The shrine welcomes all, even those on the margins of society: prostitutes, drug-dealers, petty criminals. This is not advertised, of course, but in the rambunctious, celebratory atmo- sphere of the Urs, there is an unmistakable air of quiet acceptance, as people from all different social groups and classes come together.

This is quite an established tradition at the centuries-old dargah. A Mughal chronicler, describing Qutb Sahib's Urs, notes that

> In every lane and street pleasure seekers search for carnal
> pleasures and dance with joy. Those who drink do so without
> worrying about the public censor.... The bazaars and lanes are
> crowded with aristocrats and nobles and every nook and corner
> is abuzz with the rich and the poor.[15]

As this passage suggests, despite the permissive atmosphere, there were still stark status distinctions. This remains true even today. The ordinary devotees generally sit on the ground in front of the qawwali performers. The community notables, on the other hand, are elevated above the crowd, sitting on a wooden platform, dressed resplendently. Their power is both spiritual and secular; sitting on the stage are those who lead the management of the shrine and who continue to refine its devotional practices. They maintain close connections with the leaders of other shrines, to whom they are bound by ties of belief and, more practically, of marriage.

Seated behind the notables are a more raucous crowd, who have given up the ties of family and have devoted themselves to the ascetic path. However, being a Sufi mystic, just like being a Nath yogi, does not involve giving up the pleasures of the world, but spiritualizing them. There is an ecstatic love of the divine, a love aided by quite literal intoxication. When I attended the Urs celebrations in 2016, the mystic side of the crowd was ensconced in a haze of marijuana smoke. The renunciates smoked from chillums and rolled joints to be passed around the mystic crowd.

These rollicking mystics had little patience for the more scholastic Sufis, including some of the notables. A verbal fight broke out, which threatened to become physical, when a senior spiritual leader of the shrine, the author of several books on Sufism, asked the qawwali performers only to sing songs in honor of Qutb Sahib. The request was part of an ongoing tussle between Sufism and more orthodox forms of Islam, which portray Sufis as promiscuous in their worshiping of different saints at different shrines. Only invoking Qutb Sahib's name in the saint's shrine was one way of obliquely responding to this criticism.

This irked many of the mystics, who had traveled to Mehrauli from many different shrines, and anyway spent little time worrying about orthodox criticisms of their behavior. Several stood up and started shouting when the performers weren't allowed to sing paeans to other saints, and a crowd soon formed around the two sides. The qawwali singers, seasoned players, said some soothing words and quickly launched into a new song, which defused the tensions and sent the mystics back to their joints and their chillums, and their swaying, swooning reveries.

Stoner Sadhus

Love of marijuana is yet another commonality linking the Sufis to the yogis. In many of the tantric texts, the virtues of the intoxicating plant are extolled. One text avers that marijuana is essential to ecstasy.

The plant is referred to as "victory" and "Gorakhnath's root".[16] And, as Sufis gather at Qutb Sahib's shrine to smoke, sway and (occasionally) scream and shout, groups of Nath Siddhas convene close by, on the northern edges of Sanjay Van, where three Gorakhnath Mandirs have been erected.

One of these temples, by far the biggest, adjoins the main road and regularly holds large gatherings, culminating in a biannual mela that draws significant crowds. The smallest of the temples, by contrast, is just a low brick wall surrounding several idols, protected by a solitary priest who sleeps beside the temple in a makeshift tent. The third temple combines the remoteness of the small mandir with the sociality of the big one. It is set back, away from the paved roads, in the midst of the jungle of Sanjay Van. It houses a small community of Nath yogis, who receive regular visits from devout Hindus residing in the nearby neighborhoods.

The third temple is a relatively new structure. It likely dates back only to the 1970s and many of the additions are much newer than that. But its founding, like many structures on the Ridge, from the Hauz-i-Shamsi to Bela's tomb, was divinely ordained. This, at least, is what I was told by two men from Katwaria Sarai, the neighborhood directly north of Sanjay Van. Not long ago, they said, a sadhu had a dream of Gorakhnath himself appearing in the forest. The immortal sage asked the sadhu to dig at the spot where he was standing. When the sadhu located the spot and dug as instructed, he found several clay idols. The temple was erected on the spot to commemorate the miracle.

This temple happened to be situated along my commuting route when I would walk through Sanjay Van, from Mehrauli on one side to the Delhi State Archives on the other. I spent many afternoons there, talking with the sadhus and their guests. They were as curious about me as I was about them. They invited me into their smoking circle, as they packed chillums and rolled joints. Once, when visiting with a friend, we asked where they got their supplies from. At first, they didn't quite understand the question, or, more accurately, they thought that the answer was so obvious that they didn't understand why we had asked. Eventually, they just gestured in front of them, and there, growing in the lawn outside the temple, were several marijuana plants. It seems that the barrenness of the Ridge's soil was not enough to keep the miraculous victory plant from taking root.

Marijuana plants aside, the temple always struck me as an idyllic place. The sadhus spent their time playing with a small white dog named Rocky and a big black dog named Julie, along with a small litter of unnamed puppies. They cooked hearty meals over an open flame. During one of my visits, a sadhu was hard at work preparing lemon pickle. Another day, the sadhus were crowded around listening to Shiva chants on an old CD player.

I am hardly the first to note the playfulness and lighthearted-ness of sadhus. In the late seventeenth century, the French traveler Francois Bernier noted that yogis are those who "scoff at every-thing, and whom nothing troubles".[17] The yogis think of themselves in this way too. In one text from the Nath tradition, this attitude is summed up:

> Whose friend is a Yogi when he plays? It takes so little to please him. He doesn't give a thought to what's high or what's low. Whatever he wants to do, he just does it... When you're carefree you want for nothing. From a pauper to a king, from a king to a pauper, never bothered over the difference between the two.[18]

But for all their playfulness, life with the sadhus is not a total free-for-all. At the Sanjay Van temple, as elsewhere, there is a strict hierar-chy amongst the sadhus. The youngest of the sadhus, who never told me his age but who looked to be in his late teens, was constantly being sent on errands. Older sadhus would throw crumpled banknotes at him and order him to go to nearby Katwaria Sarai to pick up milk for tea, or to buy vegetables for the day's dinner. The senior sadhu, who acted as the head of the temple, was revered, and always was the first to be greeted and bowed to. Once, as a token of appreciation for the sadhus' hospitality, I had brought them a pre-rolled joint. They were quite pleased with the gift, and insisted that I light the joint, and then immediately give it as an offering to the senior sadhu, who accepted it with a serenity that bordered on indifference.

Despite their apparently free-spirited life, their tattered clothes and their copious marijuana use, the sadhus were treated with the utmost respect by the nearby residents and, perhaps more surprisingly, by the immaculately-dressed pujaris who occasionally came to the temple to perform rituals. It was an odd sight. The sadhus spent hours smoking up, playing with puppies and drinking chai, and yet, they commanded the unwavering respect and devotion of those who were, by normal standards of society, far more upright and industrious. Residents often came bearing gifts of vegetables or money, and the head sadhu kept a thick register book to keep track of all the transactions.

The deference accorded to the unkempt sadhus, though surprising at first glance, confirms the picture of Nath yogis that has been passed down through generations of legends and folklore. Precisely because they are renegades who care little for social norms, they evoke a mixture of fear and wonder. They work miracles but keep their secrets hidden behind a cloud of marijuana smoke and arcane ritual.

With those who know the sadhus better, however, the general attitude of fear and respect has its limits. There are several local residents who have bonded with the sadhus, smoking beedis and joints with them, wiling away summer afternoons and exchanging small talk. These guests are also drawn into the informal economy of the temple, running errands and helping organize the donations that come pouring in. Once, as I sat smoking at the temple, one of these residents, following the instructions of a sadhu, took a 500-rupee note from a construction worker who was temporarily living in the temple, putted off on his scooter, and returned a half hour later with a bottle of whiskey and some change, both of which he handed to the construction worker.

His duties discharged, the man turned to the circle of sadhus and saw, next to them, a litter of puppies suckling at the teat of their mother. The man was dismayed and shouted at the sadhus for letting such an impure practice take place at the doorstep of a temple. He then turned to the dogs, shouted at them, and drove the mother away from the temple. Still incensed, the man got into a fight with one of the older sadhus, the details of which I could not catch, but whose main content was clearly related to the dogs. The sadhus all gathered around, but playful smiles danced on their faces, and soon enough, the man had calmed down and sat down for a smoke.

This was the only time I saw a resident question the sadhus and puncture the aura of respect that surrounded them. It's significant that this fight took place over an issue of purity and pollution, and specifically, the polluting influence of maternal fluids (even if they're just a dog's). Though their transgressive ways add to the yogis' mystique and enhance their image in the popular imagination, these very practices can also lead to their ostracization if they overstep societal bounds one too many times.

PART II: SECULAR SPIRITS

Breaking the Rules on the Ridge

As the Sanjay Van stories suggest, Sufis and Nath yogis are not the only ones pushing the boundaries of "civilized" behavior on the Ridge. At Qutb Sahib's shrine and at Gorakhnath mandirs, they are joined by local residents and other visitors in their pot-smoking sessions. Away from these holy sites, hidden in densely forested parts of Sanjay Van, there are signs of enjoyment that seem far from religious: empty Old Monk rum bottles, discarded condoms, cigarette butts, torn playing cards. The forest becomes a place of transgression, a place to pursue taboo activities and forbidden loves.

These perceptions and uses of the forest have quite a long lineage. In the *Mahabharata*, for instance, Arjuna was familiar with the wild side of the forest, and, for him, it was not entirely a bad thing, despite its seemingly dire, "impure" reputation. He was drawn to its adventure, its danger, its mysteries. In fact, he seemed to prefer it to the responsibilities and rigors of courtly life. After moving to Indraprastha, but before setting fire to the Khandava Forest, Arjuna voluntarily exiled himself to the surrounding wilderness.

During his exile, within three pages of the *Mahabharata*, Arjuna has had three different encounters with mysterious beauties.[19] Despite their sexual nature, these are presented as auspicious meetings, in accord with the laws of Dharma, condoned by the seers, sages and bards who accompany Arjuna in his forest exile.

First, he spends the night with a seductive snake princess named Ulupi, who brings him to her father's underwater palace and confesses

The 'Nature' of Hierarchy

her love for him. Arjuna then heads east and meets Chitrangada, the "buxom daughter" of the "law-minded king of Manalura" and "desired her". The King hears of his desires and replies, "So let her bring forth a son, who shall be the dynast; this son I demand as my price for her. By this covenant you must take her." After his tryst with Chitrangada, Arjuna heads to the "fords of the southern ocean, very sacred all and ornamented with aesthetics". But the fords are inhabited by five vicious crocodiles, who have been preying on local sages. Arjuna takes it as his duty to defeat the crocodiles in battle and jumps into the ford. Arjuna wrestles with the first crocodile he finds, pulls it out of the water, and lo and behold, it turns into "a beautiful woman decked with all the ornaments, fairly blazing with beauty, celestial and wonderful".

This is just a small sampling from the epic that famously proclaims, "whatever is found here may be found somewhere else, but what is not found here is found nowhere!" Even this three-page fragment is enough to show the tensions that the forest carried, with the text's desperate, often post-facto attempts to justify Arjuna's wild behavior and his predilection for the temptations of the forest and of the flesh.

Despite the effort to tame it and to make it submit to the codes of Dharma, the forest is nonetheless a place to find something transgressive, something beyond the pale of normal society and the structures and strictures of civilization. This makes the forest a place of pleasure, but also of danger. For every snake princess, there is a real snake, ready to attack.

What's remarkable is how much these ancient views of the forest, and these ancient tensions, still hold true today in the Delhi Ridge. The present-day Ridge forest may be of quite recent vintage. It may be filled with Mexican mesquites and ornamental flowers. The new trees of the Ridge may obscure centuries of pastoral use and state dispossession. And yet, in patches, the Ridge now undeniably exists as a forest, and it is used by an admirably wide swath of the population for activities that would not be unfamiliar to the bards of ancient India, even if there are many modern twists and surprises. The forest is still seen as a realm outside of civilization, with all the good and all the bad this implies. But in the end, on the Delhi Ridge (as elsewhere), the strict separation of the two eventually breaks down;

the realities of "civilized" life impinge on the idylls and the dangers of the forest, and vice-versa.

Intoxications on the Ridge

Like Arjuna, the young lovers in Kamla Nehru Park, seen in the introduction of this chapter, experience the forest as an erotically charged space. Their escape just happens to be in the middle of the city, not in far-off mountains and fords. And unlike Arjuna, they don't have the privilege of being backed by a brigade of Brahmins rationalizing their behavior. Instead, they get scolded by moralizing pujaris.

However, there are zones of the Ridge where a truce has been reached between the guardians of morality and the romantically inclined. People seem to understand the necessity of an escape valve for the sexual energy of the city's youth. For students living in single-sex hostels, or cooped up with their parents, there are not many outlets for romance. Delhi's proliferating malls have become a favorite spot for couples (here, apparently, public affection is accepted as part of the Western package), but these spaces are out of reach for wide swaths of the population. A much more diverse cross-section of couples can be seen in Delhi's parks.

The very publicness of the parks is what often makes it so appealing for some of the young women who frequent the Ridge. Sitting on a bench along a well-traveled path, within earshot of guards, there is a guarantee that things won't go too far. As women balance their own desires with both the pressures from their boyfriends and the insistent judgments of society, the parks offer a measure of anonymity, but not total isolation.[20]

But then, the Ridge is both park and forest; indeed, the tension over these two uses has long been a source of conflict between wilderness-orientated environmentalists and horticulture-oriented government agencies. These two different ecological visions of the Ridge also turn out to produce very different moral and emotional landscapes. If the parks of the Ridge are sites of measured romance, of pushing, but not too hard, against moral norms, then the forests of the Ridge are something more transgressive: passionate, hidden, unrestrained, wild, but also risky, dangerous, perilous.

Forbidden love thus finds its way to the Ridge. This sometimes leads to tragedy, especially when the contradictions between idealized love and social convention are too great. The Ridge transforms from a temporary escape to a final retreat from the crushing realities of life. In 2013, two lovers went into the depths of the Northern Ridge and committed suicide together, overdosing on sleeping pills. They were distraught that their families would not accept their relationship. History repeated itself a year later, as another young couple was found dead in the Northern Ridge. Their bodies were discovered hanging from a *vilayati kikar* tree.[21]

According to local legend, these suicides have historical precedents. Buried at the nearby Lothian Cemetery is an English civilian who came to Delhi in colonial times and fell in love with an Indian woman. His lover, though, could not fight the weight of tradition. She was married off to an Indian. Upon learning of the marriage, the young man shot himself. The young man, it is said, still haunts the cemetery and its surroundings.

Not all cases of forbidden love are so tragic, nor so haunting, but all carry an undercurrent of pathos. This is particularly true of men seeking liaisons with other men in the Ridge. Many of these men don't identify as gay; some of them have never even heard the term. The idea of a gay identity has been imported from the West, along with malls. This is not to say that same-sex couplings were not a part of Indian tradition—they clearly were, as many texts and works of art attest. But there was a different vocabulary, a different, more fluid set of norms, distinct from the current conception of being gay.

Under the weight of a Victorian morality imposed by the British and a Brahminical sense of moral purity, these indigenous traditions have largely been submerged, though they still have a subterranean existence, and they still bubble up in unexpected ways. Until the Supreme Court ruled it unconstitutional in 2018, the notorious Section 377 of the Indian Penal Code, yet another colonial law, outlawed "carnal intercourse against the order of nature", under threat of "imprisonment for life".

Such antiquated laws are not usually enforced. More typically, they have been used by the police as a means of harassment, to threaten people and collect bribes. This hardly stopped same-sex relations in India, but it has driven them to places like the Ridge. This is something I found out quite unexpectedly, when, during one of my early explorations of the Ridge, on a warm summer afternoon, I was propositioned by a man who appeared like an apparition out of the shade of the forest.

This was my sole first-hand experience with the Ridge's same-sex cruising grounds. These places, after all, don't advertise their existence, as the draw of the Ridge is largely its secrecy. But a remarkable portrait of one of these spaces has been sketched by Jeremy Seabrook, a British journalist.[22] Seabrook gives snippets of extended interviews he has conducted over many months with men in a place he simply refers to as "The Park" in an effort to protect their privacy. But there's little doubt that the Park is somewhere on the Ridge; it shows telltale signs of both its geology (large, crumbling red rocks) and its ecology (scrub and thorn forest).

The men in the Park make up a telling cross-section of present-day Delhi. In keeping with Delhi's centuries-old association with state power, many of the men are military personnel, policemen or government employees. In a nod to Delhi's new role as a center of high-end consumption, a sizable proportion of the men work in nearby five-star hotels. One of the men works as a security guard for a Japanese-owned factory in Gurgaon. Still others are in management and finance, enjoying the fruits of upward mobility and an expanding corporate sector. But many of the men are testaments to the flipside of the neoliberal dream; they have come to Delhi to escape rural crisis and oppression, but find themselves stuck in dispiriting, draining jobs. The Park is one of few places in Delhi where people on different sides of the class divide mingle unabashedly. The site also breaks down older barriers of caste and religion, as Dalits and Brahmins, Hindus and Muslims meet for trysts.

It is appropriate that such an unusual co-mingling would happen on the Ridge, that escape from staid societal norms. Men are drawn to the Park on the Ridge precisely because it is an atypical place. But calling it a park is slightly misleading. Like much of the Ridge, it has the dual character of park and forest. The latter, in local parlance, is called *jangal;* this is where the English word "jungle" comes from, and in Hindi, it carries the same connotation of wildness, ev   arity. The men may have their initial meetings on benches in   k-like section of the Ridge, but for their sexual activities, the   t to the shade of the jungle.

The forest thus plays its age-old role as a place of heightened pleasure and unusual encounters. The men consider the Ridge to be a refuge, a haven, an almost dream-like escape. Seabrook calls it "the site of a holiday from heterosexuality".[23] The adventures in the Ridge are in poignant contrast to what the men see as their familial duties: to marry, to have children, to land a reliable, well-paying job. One man uses the metaphor of day and night: the Ridge is where people come to live out their dreams, their fantasies, their unspoken desires, which dissolve with the harsh light of day, and all the responsibilities the day entails.

This is the function of the jungle. At first, it seems paradoxical that this jungle should be in the midst of a crowded city. But for many of the men, especially those from rural backgrounds, Delhi as a whole plays the same role that the Ridge plays in miniature. It is an unusual place, a place for the unexpected, for hidden loves and lusts and dangers. The alluring, disorienting role of the big city has long been a theme in Bollywood. For a country that, in the popular imagination, is largely rural (and still, for the moment, has more people in the countryside than in cities), the village is idealized as the heart of Indian civilization, the homeland of traditional values. But these values, this civilization, can also be constricting. Then the city becomes an escape, just as the forest was in the times of the epic. With the Delhi Ridge, the forest and the city are collapsed into one.

But this urban wilderness is no perfect utopia. Many of the men rue the shortness and brusqueness of encounters on the Ridge. They want to linger with a lover in the privacy of a bedroom, an opportunity few of them have. There are more immediate concerns too, as the police regularly raid the park. They patrol the jungle paths, looking for men with their pants down, arresting those that they find or letting them off after they pay a bribe. But one policeman, who frequents the park as a lover, not a raider, argues that these police actions are not just acts of simple repression; rather, the police themselves are battling with repressed desire. He says,

They will pretend they do it because they despise them, but really they want to enjoy also. If they arrest them, they will get free sex and at the same time say they are doing their duty. This is very convenient. I do not blame them. Who will not abuse the power he has, if he has the opportunity?[24]

Besides opportunistic cops, the men face more subtle dangers as well, which threaten the necessarily shadowy existence of the Park and other places like it. Another covert cruising ground, on the Northern Ridge, is located near the residence of Delhi's Chief Minister, which at the moment, is occupied by Arvind Kejriwal, leader of the Aam Aadmi Party. Kejriwal rose to fame as an anti-corruption crusader, and he often invokes the rhetoric of transparency, of shining the light on dark secrets to expose crime and shady dealings. Sometimes he means this quite literally. One of his main plans for reducing crime in the city, especially crimes against women, is by placing bright LED streetlights throughout the city, along with CCTV cameras.

This has had some unintended consequences. One is that couples looking for privacy, whether same sex or opposite sex, have fewer places to escape the gaze of a judgmental society. The new lights on the Ridge near Kejriwal's residence, for instance, have rendered it an ineffectual cruising ground. The city needs a place to keep its secrets, but, for now, the government seems intent on taking the wildness out of the Ridge wilderness.

Sexual couplings are hardly the only frowned-upon activities that take place in the Ridge. It is, for instance, a favorite place for drinking. Against the metaphorical intoxication of love, this is a much more literal intoxication, as people drink to escape their daily routine, to forget about the indignities and exhaustions of work, to cover the sharp realities of life with a fuzzy haze. Again, it is used for people who can't do such things elsewhere, who—because of social norms or parental pressures—can't drink at their leisure at home, or—because of economic compunctions—can't afford pubs or bars.

This leads to an odd mix of people who use the Ridge as a favored place for alcohol consumption. On one end of the spectrum, there are students from the super-elite embassy schools near the Central Ridge, who throw wild parties in the Ridge at night (after bribing

the guards, of course), escaping from the strictness   el or
home life. On the other end of the spectrum are elde   orking-
class men who come to the Ridge to drink country liquor and smoke
cheap beedis, whose gatherings on the Ridge are much less lavish
and much less loud than the students', but are a more regular, more
sedate affair.

Somewhere in the middle of the spectrum are the college students
who frequent the Ridge for binge-drinking. In a mix of competi-
tiveness, machismo and thrill-seeking, young men (and it is almost
entirely men, at least in the Ridge) drink themselves into oblivion.
For the most part, the consequences of this are fairly harmless: a
pounding headache the next day, some scratches and bruises from
stumbling through thorny thickets of *vilayati kikar*.

But there have been more tragic cases. In 2003, in the Northern
Ridge, a group of male students from the nearby Delhi University
were out drinking and carousing. They reached one of the Ridge's
ponds. One boy boasted to his friends that he would jump into the
deep waters of the pond and go for a swim. In his drunken playfulness,
he forgot that he did not know how to swim. Within minutes, he
drowned in the pond.

His death was attributed to drunken recklessness, but the pond
had dark connotations well before this tragic incident. Locally, it
is called "Khooni Jheel", or bloody pond. Many tie it to the bloody
events of 1857, since significant fighting took place in this area. Some
say the jheel's name refers to Indians who drowned themselves after
the British had suppressed the rebellion, as they preferred suicide to
death at the hands of the savage British. Others say that it was actually slain British troops who were dumped into the pond, and that
ghosts still haunt the area, including a headless British officer. Even
for those seeking drunken revelry and escape on the Ridge, then, the
forest, with its past and present dangers, remains a place of shadows
and sometimes morbid surprises.

Ghosts and Tragedies of a Different Sort

Even humans on the Ridge can have a haunting presence, appearing
as relics of a lost era, their past traumas echoing with present-day

suicides and deaths. This was certainly the case with the long-time inhabitants of the Central Ridge's Malcha Mahal, one of the quartzite hunting lodges built by Firoz Shah Tuglaq. Several websites describe Malcha Mahal as haunted, yet none of these sites actually provide information about ghosts or other spirits haunting the palace. Rather, they give details about the strange tale of Malcha Mahal's very human residents, who lived there between 1984 and 2018. The residents closed themselves off to the world and immersed themselves in long-shattered memories, including, they claimed, past glory as descendants of royalty.

This case long fascinated international journalists, who competed to get an audience with the reclusive family living in the palace. Very few were successful, but those who did emerged with a bizarre story that stretches back to 1856, when the British overthrew the kingdom of Awadh, roughly 250 miles from Delhi.[25] After Independence, former rulers of the "princely states" were given generous government allowances, but because the King of Awadh was deposed long before 1947, his descendants were out of luck.

In 1971, in a populist move, Indira Gandhi pushed for removing the privileges of royal descendants, arguing that a democratic country did not need such monarchical relics. After this, a woman called Begum Vilayati Mahal, claiming she was a descendant of the Awadh royal family, decided to protest Indira's anti-royal-privilege policies (despite the fact that the Awadhi family never received these privileges). She did so by occupying the first-class waiting room of New Delhi Railway Station, where she lived for ten years, demanding that the government return her ancestral properties. She was joined by her two young children (a son and a daughter) and a host of servants.

She was not thrown out because she guarded herself with eleven Doberman Pinschers and vowed to drink snake venom if anyone tried to remove her. In 1984, shortly before Indira Gandhi's death, Begum Vilayati Mahal had an impromptu meeting with the prime minister. Indira, her populist phase long over, sympathized with the begum: they were both, after all, inheritors of a dynasty, and they were both strong, defiant women. Indira promised to find a solution for the princess.

But Indira was assassinated before she could fol   rough on her vow. Her successors settled on a compromise,   gh the terms were rather unfavorable for the princess: she   not be given any of the ancestral homes in Awadh, but she could take up residence in Malcha Mahal, which now lay in ruins in the thorny underbrush of Delhi's Central Ridge. The building had no windows, no doors even, and certainly no electricity or water connections. But the princess, stir crazy after a decade in the railway station waiting room, agreed to the deal, though she vowed to keep fighting for her ancestral properties.[26]

Thus old royalty settled into an even older palace. From their new home, they could gaze out on the future; Malcha Mahal sits opposite a high-tech ground station run by the Indian Space Research Organisation (ISRO). Initially, staff members at the ISRO were sympathetic to the eccentric family and helped them set up water and electricity connections. However, Begum Vilayati Mahal's son, Ali Raza, broke the streetlamp that the ISRO staffers had set up outside the palace, since he worried that it would attract the attention of the rowdy young men who frequented the Ridge at night. As punishment for this, the staffers cut off the electricity connection, and the palace was once more plunged into darkness.

Over the decades, the family turned increasingly inward. A sign hangs over the entrance, littered with typos, but with its message nonetheless quite clear: "Entry Restricted. Cautious of Hound Dogs. Proclamation. Intruders Shall Be Gundown." Ali Raza was known to threaten journalists and other thrill-seekers with a gun. But it seems he was doing this more out of weariness than malice. He may have owned hound dogs, but he was the one who felt hounded by media sources that couldn't get enough of this story.

In the 1990s, tragedy struck. Begum Vilayat Mahal, tired of her fruitless fight against the government, committed suicide, reportedly by swallowing crushed diamonds. The children, now grown up, buried their mother behind the palace. But people started to appear on their property at night. Several young men attempted to dig up the body of the begum, convinced that she would be wearing expensive jewelry. The prince and princess, harried, dug up their mother's body and burned it.

The prince and princess stayed on in Malcha Mahal. They reported that much of their property had been stolen, including a silver table and several pieces of gold cutlery. While Ali Raza generally had testy, terse interactions with journalists, he was not initially a total recluse. His closest contact, it seems, was a Sufi mystic named Ali Khan, who spent his days wandering through the Ridge. The two would often meet at a nearby dargah. But as Ali Raza got older, he started cutting off even from Ali Khan. He became even more elusive than the renunciates of the Ridge. His sister passed away. And, in early 2018, staffers at the ISRO found him dead on the floor of his crumbling abode.

But the final twist in the saga came in late 2019, when the *New York Times* reporter Ellen Barry filed a stunning story revealing that the family's background was more complicated than it seemed.[27] The claims to royalty were, it seemed, a smokescreen, obscuring a story of violence, forced migration and delusions of grandeur. The story deserves to be read in full, but Barry suggests that the real tragedy of the family was Partition, whose effects caused Begum Vilayat Mahal to become unhinged, and whose ghosts haunt not just Qutb Sahib's shrine and other sites of Delhi violence, but the entire subcontinent.

When Wildness Becomes Brutality

The deaths on the Ridge, from star-crossed lovers to drunken swimmers to harried "princesses", reveals a darker side of the Ridge and an essential truth of the forest: escaping from civilization can be liberating, but it can also be terrifying. The literal and metaphorical darkness of the Ridge is a double-edged sword. It lets people find hidden love and seek transcendence in both mundane and profound ways, but it also hides violent passions and demonic forces. Delhi has a reputation as an unsafe city, and the Ridge is seen as a particularly dangerous zone, a shadowy space of physical and sexual violence.

In post-Independence Delhi, the Ridge first entered the consciousness of the city's middle class not due to environmental concerns, but because of a horrific crime.[28] On 26 August 1978, two teenagers attempted to hitch a ride across the city. The kids were siblings, Geeta Chopra (age 16) and her younger brother Sanjay (age 14). Their father was a naval officer, and the family lived in the military cantonment on the western slopes of the Ridge. The kids needed to travel from their home to the All India Radio office in central Delhi. They found a willing driver near their house, but he only took them as far as Gol Dak Khana, a major post office. Their route traversed the lonely road through the Central Ridge; a small part of this Ridge zone had recently been converted into Buddha Jayanti Park, but the rest remained wild and isolated.

When the siblings tried to get a second lift, from Gol Dak Khana to All India Radio, they quickly realized they were in trouble. After getting into the car of two men who offered them a ride, they saw that the inside door handles had been removed. They were trapped. They started to fight with the two men, first verbally and then physically. Several passersby, including motorists and cyclists, saw the siblings struggling with the men in the front seat, trying to break free from their grasp and escape from the vehicle. The men, though, were driving recklessly, careening through the streets and jumping red lights. None of the witnesses were able to keep up, though two noted the license plate number and reported it to the police. The bureaucratic inertia that has long plagued Delhi had dire consequences in this case, as the police at Rajendra Nagar station decided that the criminals were not in their jurisdiction.[29]

Later, it was revealed that the two men had come to Delhi from Mumbai, where they had already established themselves as car thieves and kidnappers, and that they were on the run from the Mumbai police. The men, who went by the aliases Billa and Ranga, initially intended to kidnap the siblings and hold them for ransom. When they learned that the children's father was a naval officer, they realized, first, that the family might not be as rich as they had suspected, and second, that the father might confront them instead of paying the ransom.

Panicking, they drove the car into the Central Ridge and stopped the car in the parking lot of Buddha Jayanti Park. They then dragged the siblings to the edge of the park and into the wilds of the Ridge. Both of the teenagers fought back—the men later went to the hospital to get their injuries treated, and Billa had a serious head wound— but they were overpowered. Geeta was raped and both siblings were stabbed to death with kirpans.

The case was quickly picked up by the media, which explored every angle of the gruesome crime. News reports emphasized the wildness of the Ridge, its unruliness and danger. This tragic event was a catalyst for turning the forest of the Ridge into a park, which could be used as a safe space for morning walks and yoga and other respectable pursuits.

This, however, was an incomplete, uneven process, complicated by the many political and economic factors discussed in previous chapters. The wildness of the Ridge, though tamed in some areas, was not completely suppressed. From lovers to drinkers, people still seek out the wilder sides of the Ridge, which also means that its darker aspects has not disappeared. On occult websites, people tell stories of a ghost emerging from the forests of the Central Ridge. She is the archetypal Indian ghost: a woman in a white sari. But the rumors surrounding her have disturbing echoes of the Billa-Ranga case. She is, it is said, a hitchhiker who got lost in the forest and starved to death. She re-emerged as a ghost who tries to hitch rides from cars passing by. If the vehicles refuse to stop, she chases after them, easily keeping pace with even the rashest drivers.

The untamed image of the Ridge has also made it into fiction, with a short story called "Last In, First Out", appearing in the anthology *Delhi Noir* in 2009.[30] The narrator of the story is an auto-rickshaw driver who plies his trade around Delhi University. He is driving by Kamla Nehru Park one night and stops, thinking he may be able to pick up a passenger who fears traversing the Ridge at night. He hears a tube-light breaking inside the park. Driving into the park to investigate, he finds a young man, head gushing blood, and then the young man's girlfriend, who has been raped. The narrator helps the couple as much as he can, comforting them and taking them to Hindu Rao Hospital nearby.

This being a *noir* story, the narrator then refashions himself as an informal detective. He vows to track down those responsible for this horrific crime. He eventually does. It's a pair of men, just as in the Billa-Ranga case. The narrator gives them names that suggest the beastly cruelties of the forest: Mongoose and Cobra. Mongoose claims that he has been coerced into helping Cobra commit his crimes (in real life, Ranga made similar claims). Mongoose also explains

Cobra's motive: "He says they need to be taught a lesson. They keep coming here and polluting the morals of the nation."[31]

This is a chilling extension of the pujari's logic that young lovers are polluting the Ridge. If the forest is outside the bounds of civilization, then the punishments meted out on the Ridge can also go beyond civilized notions of justice, into the realm of cruelty and twisted violence. Trying to escape the prison of society, young lovers find themselves facing something far more horrific: a shadowy vigilante justice, a dark torture chamber that enforces its moral code in barbarous, terrifying ways.

Crime meets Ecology

Both in real life and in fiction, the Ridge has been the scene of traumatic, violent crimes, which represent a disturbing reversal of the sexual freedom that the Ridge forest promises. Sometimes, though, criminal phenomena on the Ridge, and in Delhi more generally, have been less brutal, and more mysterious in nature, more spectral, more haunting.

In the summer of 2001, a beast that the newspapers dubbed "The Monkeyman" started terrorizing the city. The Monkeyman only struck once in the Ridge, in a working-class settlement called Sangam Vihar, which sits opposite the northern boundary of the Asola Bhatti Wildlife Sanctuary. But, as his name implies, the beast was a hallucinatory take on the iconic Ridge mammal: the rhesus macaque. Descriptions of the monkey man varied wildly; in early reports, he was simply a "monkey-like shadowy figure".[32] Most of these reports came from areas with high monkey populations, where it was not uncommon for monkeys to attack people who were sleeping on their roofs. Over time, as his notoriety spread, the Monkeyman morphed into something more like a cyborg, with glowing green eyes and a futuristic belt used for navigation.

Disparate incidents in different parts of the city were attributed to the Monkeyman, and rumors about his powers rapidly spread. He was lightning fast and could jump off high-rise buildings with grace and ease. One victim claimed that when he reached out to grab the Monkeyman, the monster turned into a cat.

Over time, the Monkeyman scare died down, but not before stirring deep fears and violent conspiracy theories. The Monkeyman episode played out like a collective fever dream, brought on by the heat of the summer, the darkness of Delhi nights and the stress of living in a densely packed city. As the Police Superintendent R. K. Chaturvedi astutely noted, all of the attacks happened in working-class and lower-middle-class neighborhoods, where buildings were crammed together and, without the luxury of air-conditioning, most slept on their roofs.[33] Further, almost all the attacks took place just after a power outage, as the neighborhoods were submerged in darkness. These neighborhoods were largely on the outskirts of Delhi, where much of the city's poor had been pushed by successive demolition and resettlement drives.

In this case, mysterious danger was emanating, not from the green jungle of the Ridge, but the urban jungle. Just as men cruising in the Park saw the city of Delhi itself as a disorienting, unpredictable place, the city's poor residents, pushed to the fringes by a capricious government, saw Delhi as wild, beastly. In the Sangam Vihar case, it's quite possible that the Monkeyman sighting was in fact a monkey escaped from the prison in the Wildlife Sanctuary. The government has tried to push the city's "undesirable" elements, whether human or simian, to the edge of the city. This violence haunts the city; Monkeyman is only its most surreal manifestation.[34]

There were other resonances in the Monkeyman scare, which reveal the fears, desires and prejudices lurking in the city's subconscious. Some of these were perhaps not unexpected. The Delhi branch of the Shiv Sena claimed that there was not just one Monkeyman, but rather 131 killer monkeys that Pakistan's Inter-Services Intelligence (ISI) had sent into Delhi to sow terror. Nativist political parties clearly could not resist the opportunity to link the wild Other attacking the city to the religious Other that has long haunted the imagination of Hindu nationalists.

But perhaps the most interesting interpretation of the Monkeyman phenomenon was put forward by the intellectual Sudhish Pachauri, who saw erotic undertones in the events: romance transferred from the Ridge to the rooftops.[35] The adjoining rooftops and terraces of densely packed neighborhoods, whether in small towns or in big cities, have long been used for furtive flirtations that threaten the stability of staid social norms, a kind of homegrown equivalent of the forest. This rooftop love is often unrequited, but sometimes leads to more than suggestive glances. In the heat of the summer night, lovers, unable to sleep, sneak from one terrace to the other, hoping to consummate their budding relationships.

Perhaps, then, the Monkeyman grew out of erotic encounters gone wrong, men who ended up on the wrong terrace or who crossed the line from flirtation to aggression. Whatever the case, the Monkeyman lives on in popular memory in Delhi. He stands as a potent symbol of the city's hidden longings and tensions, whether romantic, economic or religious. His cyborg form, half monkey, half machine, is particularly suggestive. Like so much in the city, the Monkeyman is a hybrid, part of him drawn from the mythical forests of the epics, part of him drawn from the high-tech playground of the cyberworld.

PART III: SACRED AND SECULAR DIVERGE AND CONVERGE ON THE RIDGE

Ridge, Religion, Real Estate

As the case of the Monkeyman suggests, Delhi's spirits are not ethereally transcendent beings, floating above reality. Instead, they too are bound up in economic circuits, in urban upheavals, in state-sponsored displacement and violence. This applies to both the "sacred" and the "secular" spirits on the Ridge, who have found themselves in the midst of two inescapable contemporary trends: the increasing dominance of real estate in the economy of Delhi, and the increasing polarization of Hindu and Muslim communities as the Hindu right finds itself politically ascendant.

Despite centuries of exchange and intermingling, the mystics on the Ridge have not been immune to the increasingly stark religious and economic divisions that now characterize Delhi. This is evident in the geography of Sanjay Van: the three Gorakhnath temples, along with several small Hindu shrines, are all clustered on the northern end of the park, closer to the Hindu-dominated neighborhood of Katwaria Sarai, while the dargah, several graves and Bela's mysterious tomb are located in the park's southern reaches, closer to Mehrauli.

The division between the two communities has been exacerbated by the increasingly high-stakes competition for control over land in Delhi. In a way, this is nothing new. Religious institutions have long been important landowners, and even the itinerant Sufis and yogis of the medieval Arid Zone sometimes gave up their nomadic ways so they could set up lucrative trading posts. But religious competition over real estate has taken on a distinctly modern flavor in post-Independence Delhi, as both the rich and poor scramble for space and compete for profit in an increasingly dense, chaotic city.

This has led to both intra-religious and inter-religious fights over land, as well as a proliferation of renunciate real estate agents. We've seen them make appearances in previous chapters. There was, for instance, the local priest at the Mangarbani sacred grove, who called in hired goons to beat up the Gurgaon birdwatchers. Or, at a more elite level, there's the Radha Soami Satsang Beas, which established an upscale retreat in the Southern Ridge, just beyond Sanjay Colony/Bhagirath Nagar.

In Sanjay Van, each of the three Gorakhnath temples has expanded considerably over the years. Sohail Hashmi, Delhi's popular oral historian, told me that he once encountered a pujari outside one of the temples, giving instructions to two assistants who were wielding chainsaws and cutting down wide swaths of forested land to make way for a temple expansion. The two larger temples have water and electricity connections and continue to grow both vertically and horizontally.

This kind of expansion has become so common that it is almost a running joke among Delhi's environmentalists. The temples, regardless of their age, brand themselves as "pracheen" (ancient). Hashmi recounts seeing a sign on the outskirts of Delhi, where the foundations of a new building were being laid: "Under Construction: Pracheen Hanuman Mandir". A new structure will often start, not as a full-fledged temple, but as a small set of idols under a tree. This is eventually fenced in, then made into a small building, then into a bigger one. Government officials, despite their tough talk about "encroachments", usually let such temples flourish, in part because demolishing a temple could draw the wrath of an increasingly assertive Hindu community.

However, some Hindus still see themselves as the aggrieved victims. During one of my first sojourns in Sanjay Van, on a dusty, lonely trail near the Gorakhnath temple, I met a man who was out for his afternoon walk. I had come with a more seasoned Sanjay Van chronicler, and she immediately struck up a conversation with the man about the park and his associations with it.

With very little prompting, the man launched into a tirade about the hypocrisy of the government (then Congress-run), which knowingly let Muslim shrines proliferate on the Ridge, while simultaneously cracking down on Hindu temples. The temples, he asserted, were built before the government declared this land a forest in 1980 (never mind their continuing expansion, which government officials have condoned), and the Hindu community had done an admirable job planting trees and feeding the animals of the forests. Unlike the Muslims, he said, who ate neelgai.

The problem, he averred, was not just a local one: it was a national-level plan to appease Muslims. His language got increasingly abusive as he warmed up to the theme. Rahul Gandhi is a coward. Sonia Gandhi is a whore. "Did you know," he muttered conspiratorially, "that Lalu Prasad Yadav visited this very area to encourage Muslims to keep on building encroachments?"

The man's impromptu speech was, in many ways, standard-issue Muslim-bashing. Even though countless reports have documented the prejudices, oppression and material deprivation that most Muslims in India face, Muslims are still seen as a coddled, protected population. This line of argument has been sharpened by Hindutva ideologues over the years, in parallel with historical arguments that portray the Muslims as a homogeneous block of invading fanatics.

Though the man in Sanjay Van was largely regurgitating propaganda, his invocation of figures like Sonia Gandhi and Lalu Prasad Yadav aptly shows how local spaces like the Ridge are drawn into national circuits of politics and religion. If some national political figures are supposedly appeasing Muslims, others have staked their careers on aggressive assertions of Hindu identity. One famous Hindutva firebrand is Yogi Adityanath, the Chief Minister of Uttar Pradesh and the *mahant* (head priest) of the Gorakhnath Mandir in

Gorakhpur, Uttar Pradesh, in other words, a key figure in the present-day Nath tradition. Several analyses have emphasized the extreme, often violent ways in which Adityanath has strayed from the earlier syncretism of the Nath tradition.[36]

The polarized geography of Sanjay Van is thus evidence of broader trends. The polarization continues unabated; in early 2020, reporters from the famously right-wing channel Zee News approached environmentalists who had documented encroachments in Sanjay Van, and pointedly asked only about Muslim places of worship (all of which long predate Sanjay Van's status as Reserved Forest, as opposed to the more recent Hindu temples). Yet if the Ridge exemplifies the fractures increasingly dividing society, it also provides some rays of hope, some sparks of inspiration about how these fractures can be healed.

Polarization and Resistance

While tensions simmer in Sanjay Van, more explosive confrontations have occurred in other parts of the Ridge, particularly in the former mining belt that has been a battleground for real estate developers, environmentalists and state officials ever since quarrying was banned. Near the village of Rangpuri, for instance, is a settlement called Israil Camp. It is named after the founder of the settlement, Israil Pradhan, one of the many middlemen whose fame and fortune rose with the quarrying boom in Delhi. His official job was as a driver, transporting stones, but he pursued a side-business in real estate. Using muscle power and connections, he set up a small settlement on what had earlier been the commons of Rangpuri. The settlement was initially inhabited by quarry workers and their families, who had to pay bribes to Israil and the police.

The workers were playing the precarious game familiar to many of Delhi's marginalized residents: bribes, uncertainty, the constant threat of eviction or demolition. After quarrying was banned, the neighborhood continued to grow, largely housing those who catered to the needs of nearby Vasant Kunj residents: cooks, maids, vendors, security guards, construction workers and drivers. Part of the neighborhood was regularized in the mid-2000s, a welcome relief for many who had lived in Israil Camp for decades.

But in 2014, a new danger emerged.[37] The neighborhood's Hanuman temple was taken over by members of the RSS. Residents of the settlement, both Hindu and Muslim, were initially pleased when RSS members used their influence to reduce police harassment in the neighborhood. But the RSS members soon started demanding bribes themselves; they were no saviors, but simply another set of middlemen.

The RSS leader in the area, Nagendra Upadhyay, wanted full control of the neighborhood and objected to the continued presence of a small Kali temple. The Kali temple's resident priest, Jogeshwar Pandit, complained that Upadhyay forced him to hand over all the Kali temple donations. In an interview with the journalist Neha Dixit, Pandit underlined a key difference between the temples: the Hanuman temple was only open to upper-caste Hindus, whereas the Kali temple was open to people of all castes and creeds.[38]

As the RSS attempted to impose its orthodox ideology on the settlement's Hindu community, it also began targeting Muslims. It tried to provoke fights between the two communities in a variety of ways, from organizing protests against the construction of new mosques to physically confronting Muslim street vendors. But the neighborhood's Muslims refused to take the bait, and they were largely supported by their Hindu neighbors, who had their own reasons to resent the new RSS presence.

These events must be put in a larger political context. In December 2013, the upstart Aam Aadmi Party (AAP), running largely on a platform of anti-corruption and promising to prioritize the interests of the "common man", exceeded everyone's expectations in the Delhi state elections, and formed a minority government with conditional support from Congress. Forty-nine days later, AAP chief Arvind Kejriwal announced his resignation, citing lack of support and inability to implement his agenda. The government was dissolved, and Delhi was left without a state government for a year, as AAP squabbled with the BJP and Congress about how to take things forward.

The old established parties used this hiatus to push for measures that had little democratic support, including the demolition of settlements deemed "illegal". With the legislative assembly dissolved, considerable power lay in the hands of Delhi's Lieutenant Governor, the modern-day equivalent of the British-era Chief Commissioner. In Delhi's centuries-old, ongoing struggle between local and central power, the latter temporarily had the upper hand. Taking advantage of this, the BJP lobbied the Lieutenant Governor to approve the demolition of Israil Camp and several other neighborhoods, frustrated by the residents' support of the Aam Aadmi Party and their staunch resistance to the RSS's communal polarization efforts.

On 27 May 2014, two crucial letters were sent. The first was written by RSS honcho Upadhyay, addressed to the area's former MLA, the BJP politician Satya Prakash Rana, and copied to the Lieutenant Governor. It read,

> *Jhuggis are being sold illegally at a large scale. This is carried out by people from the Bangladeshi and Muslim community... If this is not stopped immediately, a large Bangladeshi community will settle here. Since the area is close to the airport and the aeroplanes are really low while landing, there is a threat of a terror activity on the planes [sic].*[39]

The fevered Hindutva imagination is hard at work here: wherever there are Muslims, there must be Bangladeshis; wherever there are Muslim Bangladeshis, there must be terrorists.

The RSS and BJP were clearly working in concert, because on the same day, Rana sent his own letter to the Lieutenant Governor, steering clear of Upadhyay's conspiratorial tone, but essentially asking for the same thing: "Some organized groups are settling on this forest land and immediate action should be taken to remove them."[40]

The Lieutenant Governor was sympathetic to their plea and forwarded their complaint to the Forest Department. This land, after all, was technically a Reserved Forest, even if there were no trees in sight, and even if the settlement effort for notifying the forest had been delayed and postponed indefinitely. The demolition request took some time to process, but finally, on 25 November 2014, bulldozers appeared in Israil Camp, along with a handful of Forest Department officials and a phalanx of policemen. The residents were given 10 minutes to vacate their homes, and then the bulldozers leveled the entire site, destroying approximately 400 homes, rendering roughly 2,000 people homeless and crushing most of their possessions.

After the demolition, the BJP MLA flatly said that the action was meant as punishment for the intransigent residents, who had strongly supported the Aam Aadmi Party in the December 2013 elections. His previous emphasis on "forest land", in an area that had been thoroughly barren for many decades, was clearly a subterfuge, hiding both religious and political motives.

This was not an isolated case; the BJP was using the rhetoric of "clean and green" environmentalism as a complement to their strategy of religious polarization. The latter strategy had helped catapult the BJP to victory in the national elections in 2014, most notably with the stoking of religious tensions in Uttar Pradesh. In Delhi, with the state assembly dissolved and new elections looming, the BJP tried this strategy on a smaller scale, sparking communal tensions in the working-class neighborhoods like Bawana and Trilokpuri. When this didn't work, as in Israil Camp, they switched tactics and spoke of environmental preservation.

After the Israil Camp demolition, their efforts faltered. Working-class residents in nearby neighborhoods, especially those with large Muslim populations, knew that they could be next, as rumors of impending demolitions had been swirling for several months. At the same time, middle-class activists working on housing rights issues began preparing legal strategies to aid those whose neighborhoods were under threat.

When the BJP continued their offensive, their opponents were better prepared. In the last days of November, residents of Dalit Ekta Colony received a notice that their houses would be demolished the next morning. It was later revealed that this neighborhood had long been on the BJP radar; Dalit Ekta Colony was named in Rana's 27 May letter to the Lieutenant Governor, along with several other settlements on the Southern Ridge.

The colony, whose population is half Dalit and half Muslim, is another Ridge neighborhood whose residents mainly work in nearby Vasant Kunj, part of the half-hidden service sector that keeps the upscale side of the city running. The residents protested against the demolition, and a housing rights NGO filed a PIL on their behalf. The Delhi High Court ruled that the demolition should be postponed, since the DDA had not shown that the eviction was in keeping with government policies, and since no government officials had yet been punished for allowing the "encroachment". It was a significant, if temporary, victory.

The reprieve was welcomed by the Dalit Ekta Colony residents, but their anxiety about the future was mixed with anger at the BJP and nostalgia for the brief, 49-day rule of the Aam Aadmi Party, when police harassment had reduced, demolition threats had ceased and utility bills had been slashed. A journalist who interviewed colony residents in January 2015 found overwhelming support for AAP, along with disdain and resentment directed not just towards the BJP, but also Congress, a party many of them had previously supported, but which had repeatedly failed to fulfill its promises.[41]

This restive mood was not confined to Dalit Ekta Colony, nor to similar neighborhoods dotting the Southern Ridge. When state elections finally took place in Delhi, on 7 February 2015, the Aam Aadmi Party stormed to victory in a historic landslide, winning 67 of 70 seats. Many saw AAP's stunning victory as a major transition point, marking the ebbing of the wave of "clean and green", "world-class" Delhi propaganda. After its victory, one of the party's first moves was to place a temporary ban on all slum demolitions so that the party could review government procedures and revise them to ensure that all residents were treated justly. AAP's sensitivity to these issues, and the fact that it was the only party that even mentioned the Ridge in its election manifesto, also suggested that AAP could usher in a new era of management for the Ridge, one that recognized its ecological significance without relying on coercion and slum demolitions to police its boundaries.

Initially, AAP struggled to maintain the momentum generated by its stunning Delhi victory. Part of this was due to the age-old tensions between the local and central powers in the city. AAP's leaders had goaded the BJP from the beginning, and they had just scored a remarkable political victory over the party that was otherwise in ascendance nationwide. The central government, which has considerable control over Delhi, struck back at AAP, using every tool at its disposal to hamper the city government. But perhaps even more troubling for AAP were internal tensions, which led to the expulsion of key party members less than two months after their sweeping victory. One of the expelled members, Yogendra Yadav, was frequently described as the party's ideologue. He was pivotal in AAP's efforts to reach beyond its initial middle class base and to sketch out a transformative vision that would capture the energy of working-class Delhites in places like Dalit Ekta Colony. With Yadav gone, there were worries that AAP would just be a blandly "post-ideological", technocratic party. And despite grand ambitions, it struggled to expand beyond Delhi, causing some to write it off as a spent force.[42]

But requiems for AAP were premature. Despite the BJP's intense campaigning and further attempts at religious polarization, AAP won in another landslide in the 2020 Delhi assembly elections, taking 62 seats to the BJP's 8. The party remains popular in Delhi in part because it has publicly fought with the central government on the issue of slum demolitions, protesting a demolition that led to the death of an infant in the neighborhood of Shakurbasti in 2015, and taking issue with the DDA's destruction of a Ravidas temple on Ridge land in 2019.[43] The latter issue again shows the complex interplay of religion, land, class and caste, as the temple was extremely popular with Dalit worshippers, and the demolition has highlighted the hypocrisies of the Hindu nationalist project—calls for Hindu unity, but at the expense of the so-called "lower" castes. In the 2020 election, however, AAP largely steered clear of religious issues, reverting again to its image of a party of good governance, able to get things done in a pragmatic way. The full impact of this approach for the Ridge—and for Delhi as a whole—remains to be seen.

Messengers and Messiahs on the Ridge

If AAP's is a largely technocratic vision for Delhi's environment, what of other, more spiritual, visions? Delhi has no shortage of godmen claiming to work for the betterment of its environment. Leaving the Ridge briefly, there was the much-publicized "World Culture Festival" organized by holy man Sri Sri Ravi Shankar and his Art of Living Foundation, which took place on the flood plains of the Yamuna River. Sri Sri is a sadhu for the modernized, office-going middle classes, and those who aspire to that lifestyle. But his festival, which, amongst other things, claimed to clean up the Yamuna by showering it with a mysterious "enzyme" developed by a quack in Thailand,[44] was roundly criticized by environmentalists, again suggesting rifts within Delhi's middle classes. Many pointed to the hypocrisy of demolishing working-class settlements on these flood plains, only to repopulate them with the Commonwealth Games Village and the occasional religious mega-event.

On the other end of Delhi, on the Southern Ridge, another controversial holy man made his presence felt. The sage in question was Gurmeet Ram Rahim Singh, the now-disgraced head of the Haryana-based organization Dera Sacha Sauda (DSS). Several years in a row, on 15 August (also, conveniently, Independence Day), Singh organized massive tree plantation drives. Earlier drives took place near DSS headquarters in Sirsa, Haryana, but, like all ambitious leaders, Singh soon had his sights set on Delhi. In 2015, thousands of Singh's followers, largely from Haryana, descended on the Asola Bhatti Wildlife Sanctuary and embarked on a large-scale planting effort.

Like Sri Sri, part of his appeal is his orchestration of mega-events. But if Sri Sri, in his austere white clothes, appeals to those who want to show their refined sensibilities, the technicolor Singh appeals to those who dream of acquiring wealth to flaunt. Before his arrest in 2017, Singh was perhaps best known for the four movies he has written, directed, edited, produced, scored and starred in, garish showpieces that elicited the mockery of the urbane wits populating Facebook, Twitter and other social media sites. Beneath the mockery was a familiar phenomenon: disdain towards the flashiness of the nouveau riche.

Even his Ridge plantation efforts can be seen in this light. This is reforestation as a form of striving, in a world where bigger is better, where everything must be quantified, where world records show one's international prowess. The DSS website lists the relevant figures: 19 World Records, as certified by Guinness, some ostensibly charitable (including "Most Trees Planted Simultaneously at Multiple Locations in Eight Hours"), some more whimsical ("Most People Tossing Coins Simultaneously").

Mega-events like the plantation drive also have the benefit of being interactive, and of involving huge numbers of people in activities that make them feel like they are contributing to the good. And indeed, Singh's organization has contributed significantly, not just to tree plantation drives, but to relief efforts after earthquakes, cyclones and other natural disasters. These deeds are generally performed by the organization's Shah Satnam Ji Green 'S' Welfare Force Wing, which outfits its volunteers in matching beige uniforms and dispatches them around India.

In Asola Bhatti Wildlife Sanctuary, the Green 'S' Welfare Force was a formidable sight, as I witnessed in August 2015. They marched down a dirt lane in the sanctuary in a seemingly endless procession, most on foot, some in cars or tractors, all carrying plants. There was a lightness in the air, a sense of joy and camaraderie, despite the sweltering heat. Singh has an enormous following among Dalits, who feel alienated by the caste biases of other religious organizations, and who take a clear pride in contributing to an organization that sees them as equals.

I had gone to the sanctuary that day to check out a native tree plantation drive organized by a civic society group committed to promoting Delhi pride. It was led by an environmentalist who stressed the importance of reforesting the Ridge with trees that were native to the northern Aravallis. Our group, several dozen at most, was minuscule compared to the DSS Welfare Force, and we clearly came from a different social background: English-speaking, college-educated, urbane, middle to upper-class. In short, we were the kind of people who made fun of Singh's movies.

Our group leader was upset that the DSS group was not planting native trees, though he recognized that this was hardly their fault: the Forest Department provided them with saplings, and, two decades after the sanctuary's founding, they still did not have an adequate supply of native seeds. Others were less restrained in their complaints. I heard some grumbling: "They're making so much noise", leavened with a patronizing, "But they're happy". An uncomfortable, undeniable sense of "us" vs. "them" had emerged.

The Wildlife Sanctuary, with its high border fences, its strict rules for entry, and its costly recreational programs, has generally tried to keep people out of Ridge. The DSS event was a striking reversal of that: here were people from all classes and backgrounds, storming the gates of the sanctuary, bringing with them blaring music, mountains of fried snacks and boxes full of religious paraphernalia, to be sold in the sanctuary's parking lot. Delhi's environmentalists bemoan the fact that no one seems to care about the city's ecology, but when the masses do show their caring, they are viewed with suspicion: they are brainless followers, they're just carrying out the godman's order, and so on.

As it turned out, there was ample reason to be suspicious of DSS, just not for the reasons environmentalist were suggesting. In 2017, no longer able to outrun the rumors that dogged him with increasing persistence, Singh was arrested on charges of raping two of his female followers. A recent expose of Singh detailed the full extent of his crimes: bribery, murders, forced castrations, sexual assault and more.[45]

Yet despite the utter depravity and criminality of their leader, the sight of devotees filling the Wildlife Sanctuary with such bustling energy, devoting their time to the Ridge, was genuinely inspiring. It hints at the possibility of a mass mobilization for a Delhi that provides a true greenness and ecological health for all its residents, not just the privileged few.

Transforming the Ridge, Transforming the City

The DSS plantation drives are, though, just a hint. Leaving aside Singh's moral bankruptcy, the DSS approach barely scratches the surface of the problems—ecological, political, economic, social— facing the Ridge and the rest of Delhi. It offers spiritual healing as a band-aid, covering up the wounds of a traumatized society, but not addressing the root causes of societal illness. It caters, in large part, to those who have been dragged across the growing city-countryside rift, and have emerged battered and bruised.[46] A real solution has to involve closing this rift, and re-integrating Delhi's citizens into their larger, more-than-human world.

This is a tall order. It is unlikely to be achieved within the current economic system, which adds layer upon layer of separation between humans and their ecological means of sustenance, while depending on the exploitation of both nature and workers. Nor is it likely to be achieved within the current political system, which especially in Delhi is weighted towards elite interest and wary of mass involvement.

But the enormity of the task at hand need not induce a sense of indolence. There are signs of hope. On the other side of the Wildlife Sanctuary wall, for instance, the Ods of Bhagirath Nagar have been showing that there are ways forward, immediate demands that can initiate the long march towards a transformed future. The continuing saga of the Ods, including their continued collaboration with a varied set of political actors, suggests that the solution to the problems plaguing Delhi may not be spiritual in the conventional sense. But it will inevitably be animated by a spirit of struggle: of camaraderie, of ingenuity, of a near-religious faith in the possibility of a better world in the here and now.

When we last saw the Ods in Chapter 2, they were fighting to hold onto their land in the face of a sustained government onslaught. But their fight has not been merely defensive. They have made concrete, positive suggestions about ways in which their community can be integrated into the healthy functioning of the Wildlife Sanctuary. Some of these suggestions are incredibly simple, but they have not yet been implemented precisely because they overturn the established pattern of state-sponsored forest management, which sees humans living in or near forests as threats. For instance, Od villagers have, by and large, been excluded from the sanctuary's reforestation project, despite their proven skills and practical knowledge of the soil. There is much work to be done in the sanctuary, especially because it is not "pristine". The overabundance of *vilayati kikar*, for instance, could be seen as an opportunity for community forest management. Ods living nearby who need firewood would actually be doing an ecological service by cutting down *vilayati kikar* provided that indigenous plants are simultaneously encouraged to grow. But they have not been given the opportunity to do so.

In late 2014, several activists in the Od community began a campaign demanding recognition under the Forest Rights Act (FRA), a law formulated largely to address the glaring faults of the colonial-era Indian Forest Act.[47] The FRA gives legal rights to "traditional forest-dwellers", including ownership rights over the land, as well as rights to use the land for pasture, agriculture and other purposes.

Although it mainly applies to Scheduled Tribes, the FRA also provides provisions for communities who can prove that they have lived in, and depended on, a certain stretch of forest for over 75 years. Passed in 2006, the FRA was a belated recognition that many tribes have been the victims of massive land-grabs engineered by a forest-hungry state, and that tribal people can actually play a positive role in forest conservation. This suggests that the nature/culture divide that plagues environmental thinking is not just being questioned, but is also being overturned in practice, at least in some small ways.

The Ods' request for recognition under the FRA is highly unorthodox, pushing past the original intentions of the Act. For the most part, the Act was envisioned as a way of empowering tribes in dense, vast jungles far away from urban habitations. The Ods, on the other hand, are in the middle of a bustling metropolis. Further, they are not recognized as a Scheduled Tribe (though they have asked for this recognition as well), and they have not been in the area for more than 75 years.

But the Ods' campaign is inspiring precisely because it questions the rigid categories on which even progressive legislation like the FRA is based. They may not be ST, but it is hardly their fault that the state has not yet deigned to categorize them as such. And they may not have lived in Bhatti for the required period of time, but that is because their tribe is nomadic. It seems odd that a law developed as a defense of tribal life makes little room for one of the chief characteristics of that life: mobility. Such considerations make it clear that the Ods' request certainly accords with the spirit of the act, which is to recognize the rights of tribal groups that have long been marginalized by the state.

The Ods' argument is strengthened by the simple fact that their residence in Bhagirath Nagar significantly predates the arrival of the "forest" in the Bhatti Mines. Many of the Od families have been there since the 1960s and 1970s; the Wildlife Sanctuary was only extended into Bhatti in 1991, and even this was just on paper. The actual "forest" started to emerge in Bhatti only after the intervention of the army. And this "forest" is largely a weed-filled thicket, which was formerly a pockmarked mining landscape, which was formerly a pastoral grazing land, which was formerly a savanna. Logically, then, the Ods really are the original inhabitants of the so-called forest, since they were the ones living there when plantation efforts began.

In a sense, one can understand why some environmentalists might resist this logic. Once the Ridge is connected to larger trends—to quarrying throughout the Aravallis, to construction booms and real estate expansions, to migrant populations—then it becomes very difficult to control. Far easier just to draw a strict boundary around it, both temporally and spatially: from 1996 onwards, 80 square kilometers of the Ridge is Reserved Forest. Period.

But seeing the Ridge as interconnected can also be empowering, even for conservationists. If everything is interconnected, then one can start implementing positive changes anywhere, and this change has the potential to ripple outward. For instance, the Ods are the second urban tribe to use the FRA, after tribals in a national park in Bombay; if this approach is successful, it could spread to other cities, pioneering strategies for balancing use and conservation even in the most challenging of scenarios.

The Ods have submitted their FRA request to the state and have yet to receive a response. Several Od activists, though, have been heartened by the rise of the Aam Aadmi Party, whose government now controls the state-level Forest Department, as well as the Labour Department. It is too early to say if this new energy, both among activists and politicians, will lead to lasting changes in the Wildlife Sanctuary and beyond, but it is a sign that new movements are emerging, and old patterns are being challenged. India's tribes, often dismissed or stereotyped as primitive, are showing potential ways forward. This is not a backward-looking idealization of a rosy past, but tentative steps towards an uncertain future, towards healing the rift between town and country, human and nature.

There are other glimmers of hope as well. The plight of the Ods underscores the fact that questions about nature cannot be separated from questions about work and livelihoods. The long-term solution to Delhi's environmental problems can't just be banning factories and quarries, and then just relocating them to other parts of the country and continuing to consume what they produce. This just makes the production process, with its human and environmental costs, more obscured, and perhaps even more dangerous.

In Delhi, there have been many attempts to bring these costs to light, not just from elite "clean and green" environmentalism, but from the workers at the heart of "polluting" industries. Moving outside the boundaries of the Ridge, but staying within the realm of Delhi's parks, we find an instructive example: a small green space called Raja Park near the Wazirpur Industrial Area. In times of struggle and conflict, Raja Park has become a community planning and protest ground for the workers nearby. This included a major rally in 1996 to protest slum demolitions; more recently, the park has been used during major disputes over production conditions in the Industrial Area.

Industrial production first came to Wazirpur in the late 1970s. Many of its initial employees were those who had come to Delhi as construction workers for the Asian Games building boom and had remained in the city after the Games ended. Although initially, most of the factories were hosiery manufacturers, from the mid-1980s, small stainless-steel factories came to dominate the landscape. The working conditions in the steel plants were, and still are, inhumane. Highly dangerous acids are used. Carbon fumes cloud the workplace. The workers know from first-hand experience the dangers of industrial pollution. Skin diseases and tuberculosis are common, as are constantly swollen hands and lacerations.

There is a long history of labor activism in this area, as workers have sought to improve their work conditions. There were major strikes in 2012, 2013 and 2014.[48] In 2013, along with other tactics, the workers and their supporters occupied Raja Park, using it as a space for meetings, protests and much-needed rest. They repeated this action again in 2014, even setting up a community kitchen to support workers who had not been paid in many days.

This transformation of a green space into a vibrant site of community support and engagement, in a very politically charged moment, would likely be anathema to politicians like Jagmohan, or, more recently, Sheila Dixit, who longed for a peaceful, quiet, conflict-free Delhi, and who sought to achieve it by suppressing and demonizing those who did not fit the image of a "clean and green" city. But for those interested in constructing an environmentalism from the bottom up, one that links ecological and economic issues and remains committed to social justice, the transformation of Raja Park was a moment of admirable democratic assertion and spirited community solidarity.

The End

On a more day-to-day basis, this kind of democratic spirit suffuses the Ridge's more traveled parks and paths. In Sanjay Van or Kamla Nehru Park, despite the state's intentions, the Ridge is used by all castes and classes, for a bewildering array of uses—for love, for enlightenment, for recreation, for grazing, for firewood-gathering, for shortcuts and detours, for food, even for burial and cremation.

The Ridge, especially around Mehrauli, is covered in graves and dotted with crematoria. In a neighborhood with centuries of human history, the living jostle with the dead for space. Death has an odd way of bridging divides on the Ridge: between nature and culture, between Hindu and Muslim, between the spiritual and the material. No one can escape death. It is the ultimate, unavoidable return to nature: ashes to ashes, dust to dust. No matter how much we try to distract ourselves from this inevitability, through work, through play, through conspicuous consumption, we cannot avoid it. The Ridge's graveyards and cremation grounds are a reminder of a sacred truth: that death is necessary for life, that tragic endings are the required fuel for new beginnings.

On the Ridge, one generally finds Muslim graves and Hindu cremation grounds, but even these categories get blurred: yogis often prefer burial, and they join their Sufi brethren in this repose. The Awadhi "royals" on the Central Ridge chose to cremate their mother, although they are from a Muslim family, when a grave proved to be an unsafe final resting place. In Sanjay Van, a medieval Sufi lies next to Bela Chauhan, who was, depending on who you ask, either a Sufi mystic herself or an incarnation of Kali.

Perhaps these two incarnations are not so different. They are a reminder that, for all the disenchantment of nature, for all the attempts that have been made to tame nature (and those genders and caste groups associated with it), it still has a wildness and a sacred, mysterious aura. And as the Tantric tradition has suggested, this mystery is not just out there in nature, but within us, suffusing the material world that includes both humans and their environments.

The yogic mystic, rejecting the hierarchies and orthodoxies of the world, smeared in ash, meditating in graveyards, seems far distant from the striking steel-factory worker, seeking sustenance after an exhausting day of protest and confrontation. But both exist in Delhi's green spaces. Both blend a spiritual strength with a direct, material engagement with the world. Neither will appear in the latest ad campaign for a clean, green, pristine Delhi. They point to another vision for the Ridge, one full of surprises and hybrids, of struggle and serenity. A Ridge re-enchanted, its riches redistributed and enjoyed by all.

NOTES

Seeds

1. For ease of reading, I do not use diacritics when transliterating names (of both historical figures and places). Instead, I rely on the most commonly used English-language transliterations.
2. Beg, *Bahadur Shah*, 20.
3. The findings have since been updated; Delhi has dropped to sixth place on the list, but is still alarmingly polluted. See World Health Organization, "WHO Global Pollution Database".
4. Delhi Development Authority, *Master Plan 2001*, 4.
5. Beg, *Bahadur Shah*, 22.
6. Srishti and WWF-India, *Saving the Delhi Ridge*, 21.
7. van Buitenen, *The Mahabharata Book 1*, 417.
8. See especially Dove, "Dialectical History of 'Jungle'" and Karve, *Yuganta*, 146.
9. Diamond, "The Worst Mistake".
10. Graebar and Wengrow, "How to change".
11. Greabar and Wengrow, "How to change".
12. Scott, *Against the Grain*, 38.
13. Weisman, *The World Without Us*.
14. Botkin, *Discord Harmonies*.
15. Kathleen Morrison suggests that these two narratives sometimes converge, noting that an eco-romanticist strain of Indian environmental thinking posits a 'Hindu Eden' and imagines caste as an ecological niche keeping humans in balance with each other and with nature. See Morrison, "Conceiving Ecology", 44.
16. See, for instance, Press Trust of India, "Spiritual traditions", and Mishra, "Modi's Idea of India".
17. For a good summary of Jagmohan's impact on Delhi and its environment, see Pati, "Jagmohan: The Master Planner".
18. Ghertner, "Green evictions", 146.
19. Roy, *Pollution, Pushta, and Prejudices*.
20. Delhi Planning Department. "Economic Survey of Delhi 2012–3".
21. Baviskar, "Between violence and desire".
22. Christ, "Water related informal processes", 137–140.
23. Mawdsley, "India's Middle Classes and the Environment".

24. The classic analysis of the estrangement and alienation of labor under capitalism is elaborated by Karl Marx, "Economic and Philosophical Manuscripts".

25. For a similar argument, elaborated with more theoretical sophistication, see Smith, "Nature as Accumulation Strategy".

Chapter 1: Stones

1. Singh, *Ancient Delhi*, 14–20.
2. Panditi, "Signature Bridge".
3. Wiedenbeck, Goswami, and Roy, "Stabilization of Aravalli Craton".
4. Verma and Greiling. "Tectonic evolution of Aravalli".
5. Lahiri and Chakrabarti, "A preliminary report".
6. Trivedi, "On the Surface", 58.
7. Mellars et al., "Genetic and archaeological perspective".
8. Sharma, "Prehistoric Delhi", 19.
9. Kohn and Mithen, "Handaxes: Products of sexual selection?"
10. Lam, "The First Commodity: Handaxes".
11. Mellars et al., "Genetic and archaelogical perspective".
12. Mellars et al., "Genetic and archaelogical perspective".
13. Petraglia et al., "Population increase and environmental deterioration".
14. Trivedi, "On the Surface", 64.
15. According to another account, the inscription was in fact first discovered in 1966, in the process of—what else—real estate surveying and construction in the area. See Lahiri, *Ashoka in Ancient India*, 335.
16. For a broader reflection of Ashoka's use of stone inscriptions, see Lahiri, *Ashoka in Ancient India*, especially Ch 9.
17. Mann and Sehrawat, "City with a View", 561.
18. Singh, Mukherjee, and Kapoor, *New Delhi*, 97.
19. W. M. Hailey to Deputy Commissioner, Delhi, February 16, 1916, "Correspondence Regarding Quarries", Deputy Commissioner's Office [hereafter D.C.O.]. 19/1915, Delhi State Archives [hereafter D.S.A.].
20. H. M. Griffiths, Executive Engineer, 5th project Division to Superintending Engineer, March 27, 1916, "Correspondence Regarding Quarries", D.C.O. 19/1915, D.S.A.
21. Chief Engineer to Secretary of Imperial Delhi Committee, April 17, 1916, "Correspondence Regarding Quarries", D.C.O. 19/1915, D.S.A.
22. H. M. Griffiths to Deputy Commission, Delhi, May 6, 1916, "Correspondence Regarding Quarries", D.C.O. 19/1915, D.S.A.
23. Deputy Commissioner, Delhi to Chief Engineer, May 22, 1916, "Correspondence Regarding Quarries", D.C.O. 19/1915, D.S.A.

24. Deputy Commissioner, Delhi to Chief Commissioner, D̲ell͟ ͟ ͟23, 1916, "Correspondence Regarding Quarries", D.C.O. 19/19͟ ͟

25. Deputy Commissioner, Delhi to Chief Commissioner, D͟ ͟ ͟ ͟ ͟ly 13, 1916, "Correspondence Regarding Quarries", D.C.O. 19/1͟ ͟ .A.

26. Deputy Commissioner, Delhi to Chief Commissioner, ͟ ͟ ͟ ͟ ͟ly 13, 1916. "Correspondence Regarding Quarries", D.C.O. 19/1͟ ͟ .S.A.

27. This process is described in Soni, "Urban Conquest", 8͟4.

28. Talib, *Writing Labour*, 31.

29. Talib, *Writing Labour*, 24.

30. Talib, *Writing Labour*, 34–44.

31. Talib, *Writing Labour*, 216.

32. Talib, *Writing Labour*, 59.

33. Talib, *Writing Labour*, 62.

34. Talib, *Writing Labour*, 71.

35. Talib, *Writing Labour*, 62.

36. Talib, *Writing Labour*, 75.

37. Talib, *Writing Labour*, 91.

38. Talib, *Writing Labour*, 71.

39. Talib again provides a compelling analysis of this process. See Talib, *Writing Labor*, esp. Chapter 3.

40. The notification is reproduced in Sinha, *The Delhi Ridge*, 142.

41. Soni, "Urban Conquest", 86.

42. Hansen, "Albert Smith, Alpine Club".

43. Rosler, "Culture Class, Part III".

Chapter 2: Soil

1. Krishen, *Trees of Delhi*, 19.

2. For a painstaking summary of this consensus, see Witzel, "Autochthonous Aryans?"

3. Mayaram, "Pastoral Predicaments", 205. For a more speculative take, see Khari, *Jats and Gujars*, 2.

4. Chattopadhyaya, "Emergence of the Rajputs", 163.

5. Chattopadhyaya, "Emergence of the Rajputs", 166.

6. Chattopadhyaya, "Emergence of the Rajputs", 165.

7. Tripathi, *History of Kanauj*.

8. Singh, *Ancient Delhi*, 92.

9. Khari, *Jats and Gujars*, 89–90.

10. Gommans, "The silent frontier of South Asia", 4.

11. Gommans, "The silent frontier", 11.

12. Goel, *Heroic Hindu Resistance*.

13. Mayaram, "Pastoral Predicaments", 197, and Khari, *Jats and Gujars*, Introduction.

14. Khari, *Jats and Gujars*, 89–90.

15. This is not to deny the well-documented violence of the Mughals, and the brutality that is common to all empires, whether they are ostensibly Muslim, Hindu, Buddhist, Christian, secular or atheist. When the cornerstone for Shahjahanabad was laid, "the bodies of several freshly beheaded criminals were put in the trenches round the cornerstone." See Blake, *Shahjahanabad*, 30.

16. Gupta, "Delhi and its Hinterlands", 142.

17. Chakravarty-Kaul, *Common Lands*, 169.

18. Punjab Government, *Gazetteer Delhi District, 1912*, 140–1.

19. For an overview of this corrupt history, see Dirks, *The Scandal of Empire*.

20. The term "wastelands" is still used in discussions of Indian ecology, both by governmental groups and by academics and environmentalists; one government publication features maps that label the entire Southern Ridge as "wasteland"—and this several years after the zone had been named a Reserved Forest! See National Capital Region Planning Board, *Delhi 1999*.

21. For an extensive analysis, see Guha, *A Rule of Property*.

22. See Bandyopadhyay, *From Plassey to Partition*, 82–96.

23. Quoted in Chakravarty-Kaul, *Common Lands*, 67.

24. Punjab Government, *Gazetteer of Delhi, 1883–4*, 74.

25. Shail Mayaram, for instance, describes "mixed caste bandit groups including gangs of Minas, Gujars, Mewatis and occasionally, even Brahmans and often organised by Rajputs." See Mayaram, "Pastoral Predicaments", 198.

26. See, for instance, Stokes, *The Peasant and Raj*.

27. This argument goes against the commonsense imagination of savannas as grassy, treeless plains; ecologically speaking, some savannas, especially those with higher rainfall, have a fairly high density of trees. Their defining characteristic, rather, is a mix of grasses and trees that are fire tolerant and shade intolerant. See Ratnam et al., "A 'forest' a savanna?"

28. Chakravarty-Kaul, *Common Lands*, 177.

29. *Gazetteer of Delhi, 1883–4*, 2. This gazetteer was based, almost verbatim, on Machonachie's Settlement Report.

30. The rivalry continues in present-day Delhi, with Jats still holding the upper hand in terms of both population and political clout. See Khari, *Jats and Gujars*, Introduction.

31. *Gazetteer of Delhi, 1883–4*, 41.

32. *Gazetteer of Delhi, 1883–4*, 55.

33. Quoted in Mann and Sehrawat, "City with a View", 552.
34. Quoted in Chakravarty-Kaul, *Common Lands*, 169.
35. Henry Beadon, "Relationship Between Forests & the Retention of Atmospheric Moisture & Soil Moisture", D.C.O. 19/1908, D.S.A.
36. Beadon, "Relationship Between Forests", D.C.O. 19/1908, D.S.A.
37. He does, at least, show some (false?) modesty at the end of his letter, noting, "In conclusion, I must apologize for the very egotistical tone of my letter, in which I have referred to my own reports, but it just happens that I have been on special duty for four years in connection with forests and so my opportunities for observation have been unusual." See Beadon, "Relationship Between Forests", D.C.O. 19/1908, D.S.A.
38. Quoted in Mann and Sehrawat, "City with a View", 554.
39. Quoted in Mann and Sehrawat, "City with a View", 556.
40. The notification is reproduced in Sinha, *The Delhi Ridge*, 132.
41. Kumar, "Birdwatchers Thrashed".
42. This process and its consequences are described in greater detail in Baviskar, "Urban Jungles", 47–50.
43. These controversies have been explored in some depth by the journalist Chander Suta Dogra. See Dogra, "Aravallis being gobbled up", and Dogra, "Village common property".
44. It may, though, be shrinking. See Ahlawat, "Forest official's kin".
45. This section is based on field visits to Od communities in Bhagirath Nagar, Delhi, between 2011 and 2015, which also included conversations with the activist/scholar Anita Soni. I have also drawn on Soni's writing, including Soni, "Displacement Woes I", and Soni, "Displacement Woes II".
46. Singh, "Delhi: Forest dept".
47. Environmentalists have worked hard to re-alienate it, with statements like "It may seem churlish to resist popular usage, except that (a) it is confusing and (b) it is founded on ignorance." See Krishen, *Trees of Delhi*, 278–9.
48. Shrivastava, "Simians' Sanctuary".
49. Vij, "Monkey Business" and Vij, "Indian Parliament's monkey problem".
50. Richard, Goldstein and Dewar, "Weed macaques".
51. Maestripieri, *Macachiavellian Intelligence*, 33.

Chapter 3: State

1. Quoted in Fanshawe, *Delhi: Past and Present*, 222–23.
2. The following account of the pillar draws on Singh, "The Later Histories", and Hashmi, "The Lives And Times".

3. Max Weber, one of the so-called "fathers" of sociology, famously defined the state as an organization that has a monopoly of the legitimate use of violence; see Weber, "Politics as a Vocation", 33.
4. Quoted in Kumar, *Emergence of Delhi Sultanate*, 115.
5. Quoted in Kumar, *Emergence of Delhi Sultanate*, 93.
6. For more on this claim and the meaning of *Qubbat*, see Kumar, *Present in Delhi's Pasts*, 32.
7. See Kumar, *Present in Delhi's Pasts*, 41.
8. Kumar, *Emergence of the Delhi Sultanate*, 357.
9. Timur, *Malfūzāt-e Tīmūrī*, 196. This particular quote, and the ones that follow, are taken from an English translation of Malfūzāt-e Tīmūrī, a supposedly autobiographical account of Timur's life; however, scholars have long doubted whether this document was actually penned by Timur, or was cobbled together by later court chroniclers based on other accounts of Timur's life.
10. Timur, *Malfūzāt-e Tīmūrī*, 214.
11. Timur, *Malfūzāt-e Tīmūrī*, 220.
12. Romila Thapar, for instance, accepts at face value Timur's account of the raid of Delhi, and reports that Timur attacked the Sultanate because "the Tughluqs were not good Muslims, and therefore had to be punished." See Thapar, *A History of India*, 280.
13. See for instance, Chandra, *Essays on Medieval Indian History*, 125.
14. Timur, *Malfūzāt-e Tīmūrī*, 218.
15. Chandra, *Essays on Medieval Indian History*, 154.
16. Quoted in Chandra, *Essays on Medieval Indian History*, 154.
17. The centrality of Delhi to the Uprising is one of the main theses of Dalrymple, *The Last Mughal*. In my description of the events of the Uprising, I have drawn extensively from that book's meticulously documented narrative.
18. Keith Young, *Delhi—1857*, 273.
19. A fascinating examination of these intra-British tensions can be found in Stanley, *White Mutiny*.
20. Fanshawe, *Delhi: Past and Present*, xii.
21. Quoted in Sengupta, *Delhi Metropolitan*, 208.
22. The story is recounted in Beg, *Bahadur Shah*.
23. Quoted in Dalrymple, *The Last Mughal*, 419.
24. Not that the British had no role in the historical violence of Partition. Perry Anderson holds Louis Mountbatten, last Viceroy of British India, particularly to blame for insisting on a "ludicrously early date" for Partition. Says Anderson, "Having lit the fuse, Mountbatten handed over the buildings to their new owners hours before they blew up, in what

has a good claim to be the most single contemptible act in   of empire." See Anderson, "Why Partition".

25. Quoted in Anderson, "Why Partition", 18.
26. Kumar, *Present in Delhi's Pasts*, 2–52.
27. Lord Crewe, Secretary of State for India, to Lord Hardinge, November 1, 1911, National Archives of India [hereafter N.A.I], *From Ghalib's Dilli*, 2.
28. Crewe to Hardinge, N.A.I., *From Ghalib's Dilli*, 2.
29. Delhi Town Planning Committee, *First Report*, 4–5.
30. Delhi Town Planning Committee, *First Report*, 4.
31. Quoted in Bajaj, "Building of New Delhi", 63.
32. Quoted in Bajaj, "Building of New Delhi", 64.
33. George S.C. Swinton to Lord Hardinge, December 19, 1912, N.A.I. *From Ghalib's Dilli*, 66.
34. These optical considerations are emphasized in Mann and Sehrawat, "City with a View".
35. The Delhi Town Planning Committee captures this ambivalence with typical understatement: "Incidentally the land which will be acquired for this site [around Raisina Hill] is extremely cheap, and while no consideration of expense should... limit the acquisition of land... it is a fortunate circumstance that the moderate price of land in this part of the Delhi district will render it impossible to entertain even a thought of such limitation." Delhi Town Planning Committee, *First Report*, 8.
36. Delhi Town Planning Committee, *First Report*, 7.
37. See, for instance, Henry Beadon to the Commissioner of Delhi, January 19, 1912, N.A.I. *From Ghalib's Dilli*, 153.
38. The fiasco is recorded in a series of letters in "Contour maps of land to be acquired for the new capital", D.C.O. 62/1912, D.S.A.
39. Henry Beadon to Secretary, Government of India, Home Department, August 5, 1912, N.A.I. *From Ghalib's Dilli*, 183.
40. "Final Report of the Imperial Delhi Committee on the Land Acquisition Procedures connected with New Delhi." D.C.O 68/1917, D.S.A.
41. My account of the Malcha saga, including all direct quotes, draws largely from two outstanding articles on the ongoing legal battle: Parashar, "The natives strike back" and Parashar, "Still waiting".
42. Parashar, "Still waiting".
43. Mann and Sehrawat, "City with a View", 557.
44. Coventry, *Scheme for Afforestation*, 1.
45. The notification is reproduced in Sinha, *Introduction to Delhi Ridge*, 131.
46. Coventry, *Scheme for Afforestation*, 1.
47. Parker, "Afforestation of the Ridge", 25.
48. Coventry, *Scheme for Afforestation*, 3.

49. The notification is reproduced in Sinha, *Introduction to Delhi Ridge*, 132.
50. Delhi Town Planning Committee, *First Report*, 16.
51. Coventry, *Scheme for Afforestation*, 11.
52. Parker, "Afforestation of the Ridge", 22.
53. Parker, "Afforestation of the Ridge", 26.
54. Details about Mustoe's intervention in the Ridge can be found in Baviskar, *First Garden*, especially chapters 1 and 3. Particularly notable is Baviskar's assertion that Mustoe had a leading role in creating both the "wild" part of the Ridge and the exquisitely manicured gardens behind the Viceroy's House; "wilderness was cultivated as much as the rest of the landscape." See Baviskar, *First Garden*, 15.
55. The notification is reproduced in Sinha, *Introduction to Delhi Ridge*, 134.
56. This account of the DIT draws largely from Mehra, "Planning Delhi".
57. This account of the Master plans making relies on the first chapter of Sundaram, *Pirate Modernity*.
58. Mehra, "Planning Delhi", 373.
59. Delhi Development Authority, *Master Plan for Delhi*.
60. Sundaram, *Pirate Modernity*, 38.
61. Sundaram, *Pirate Modernity*, 55.
62. Delhi Development Authority, *Master Plan for Delhi*, 34. Their geography is a bit off; Hyde Park is in London. But this conflation of the US and UK underlines the main sources of inspiration for the plan.
63. Delhi Development Authority, *Master Plan for Delhi*, 34.
64. Baviskar, "Cows, Cars and Cycle-rickshaws", 397.
65. These reports are quoted in Bhan, "Planned Illegalities".
66. Jagmohan, *Shaping India's New Destiny*, 199.
67. Jagmohan, *Shaping India's New Destiny*, ii.
68. Jagmohan's fondness of Haussmann, and his overall planning philosophy, are discussed in Pati, "Jagmohan: The Master Planner".
69. Quoted in Pati, "Jagmohan: the Master Planner", 51.
70. Jagmohan, *Island of Truth* (Delhi: Vikas Publishing, 1978).
71. Jagmohan, *Shaping India's New Destiny*, 160.
72. These notifications are reproduced in Sinha, *Introduction to Delhi Ridge*, 136–7.
73. Jagmohan's notifications are reproduced in Sinha, *Introduction to Delhi Ridge*, 138–40
74. Jagmohan, *Challenge of Our Cities*, 7.
75. Baviskar, "Spectacular Events", 149.
76. Baviskar, "Spectacular Events", 152.

77. Donthi, "Under Jagmohan".
78. Jagmohan, *My Frozen Turbulence.*
79. Jagmohan, *Shaping India's New Destiny*, 140.
80. Balban's campaign is detailed in Kumar, *Emergence of Delhi Sultanate*, 333–4.
81. Coventry, *Scheme for Afforestation*, 1.
82. Kalpavriskh, *The Delhi Ridge Forest.*
83. See, for instance, the account in Baviskar, "Urban Jungles", 51.
84. The activities of Kalpavriksh and other early Ridge advocates are narrated in Agarwal, "Fight for a Forest".
85. The ever-expanding ambit of industrial closures, and the militant response to this phenomenon, are explored in Nigam, "Industrial Closures in Delhi".
86. This language can be seen in Kalpavriksh, *The Delhi Ridge Forest.*
87. For more on this groundswell of activism, see Srishti and WWF-India, *Saving the Delhi Ridge.*
88. The report is officially known as the Government of NCT of Delhi, "Report of the Committee to Recommend the Pattern of Management of the Delhi Ridge".
89. Sinha, *Introduction to Delhi Ridge*, 143.
90. Sinha, *Introduction to Delhi Ridge*, vii.
91. Sinha, *Introduction to Delhi Ridge*, 145.
92. These cases are summarized in Department of Forests and Wildlife, "Pending Court Cases".
93. Ashok, "Asola Sanctuary encroachments razed".
94. Nandi, "Revenue dept to demarcate".
95. See, for instance, Press Trust of India, "Delhi ridge" and Gandhiok, "Reclaimed land".
96. Crowley, "Demolitions in Aya Nagar".
97. Quoted in Nandi, "Revenue dept to demarcate".

Chapter 4: Surplus

1. Quoted in Dalrymple, *City of Djinns*, 109.
2. Dalrymple, *City of Djinns*, 142.
3. Dalrymple, *City of Djinns*, 139.
4. Bayley and Metcalfe, *Golden Calm*, 146.
5. Quoted in Varma and Shankar, *Mansions at Dusk*, 90.
6. For more on Metcalfe's demise see Dalrymple, *The Last Mughal*, especially chapter 4.
7. Quoted in Mehra, "Planning Delhi", 19.

8. This, and other aspects of the scandal, were exposed in great detail in a 471-page report commissioned by the Janata Party after Emergency; see Government of India, "Inquiry on Maruti Affairs".

9. See Chibber, *Locked in Place*.

10. The workers' struggle is detailed in People's Union for Democratic Rights, *Driving Force*.

11. Quoted in IANS, "K.P. Singh".

12. Delhi Development Authority, *Master Plan for Delhi*, 3.

13. See, for instance, Verma, "Haryana TCP".

14. Rao, "Black, bold and bountiful".

15. For more on this process, see Searle, "Conflict and Commensuration".

16. Kannoth and Nimkar, "Realty faces reality".

17. On the complex interplay of local and transnational capital in these dry-ups, see Searle, "The contradictions of mediation".

18. Srivastava, "DLF: The ailing giant".

19. Monga and Mangal, *DLF Limited: A Crumbling Edifice*.

20. Singh, "Behind Robert Vadra's fortune".

21. Marx, *Capital: Volume 3*, 465.

22. Kumar, "Gurgaon on its deathbed".

23. This vision is analyzed in Negi, "Neoliberalism, Environmentalism and Politics".

24. Quoted in Negi, "Neoliberalism, Environmentalism and Politics", 187.

25. Quoted in Ghertner, "Green Evictions", 149.

26. Quoted in Ghertner, "Green Evictions", 155.

27. Bobb and Gupta, *Delhi Then and Now*, 11.

28. This figure is cited in Jenkins and Malik, "Deepak Bhardwaj".

29. Press Trust of India, "Deepak Bhardwaj murder case".

30. Quoted in Vij, "The BSP Billionaires".

31. See in particular Jenkins and Malik, "Deepak Bhardwaj", and Narayan and Apurva, "Making of a Tycoon".

32. Chauhan and Ghosh, "Ponty Chadha".

33. Soni recounts his story in his book *Naturally: Tread Softly*.

34. Soni, *Naturally: Tread Softly*, Prelude.

35. Staff Reporter, "Residents term public hearings".

36. Quoted in Soni, *Naturally: Tread Softly*, Prelude.

37. Quoted in Byrnes, "Victor Gruen".

38. The mall's dwindling popularity in the U.S. is discussed in Badger, "The Shopping Mall".

39. The retailers are quoted in Express News Service, "Retailers down shutters".

40. These figures are taken from Gooptu, "Neoliberal Subjectivity, Enterprise Culture".

41. The owner is quoted in Srivastava, *Entangled Urbanism*, 222.
42. Srivastava, *Entangled Urbanism*, esp Ch. 9.
43. The publicly available EIAs, quoted extensively in this section, include: Ambience Developers Pvt. Ltd., "Environmental Impact Assessment, Ambi"; Jasmine Projects Pvt. Ltd., "Executive Summary"; Ultra Tech, "Executive Summary of Environmental Assessment"; EST Consultants Pvt. Ltd., "Environmental Impact Assessment 'Promenade'"; and EST Consultants Pvt. Ltd., "Environmental Impact Assessment 'Emporio'".
44. See Staff Reporter, "Residents term public hearings".
45. Quoted in Press Trust of India, "Developers of Ambience Mall".
46. For more on this sanitation paranoia, see Mann, "Delhi's Belly".
47. "Acquisition of a Plot of Land Between Paharganj & Kutab Road for Staff Quarters," D.C.O. 11/1903, D.S.A.
48. The following exchanges are taken from "Transfer to the Civil Authorities of Land Lying South of the New Road Through Hindu Rao Estate", D.C.O. 8/1872, D.S.A.
49. This incident is recounted in Baviskar, "Between violence and desire".
50. See, for instance, Press Trust of India, "NGT prohibits dumping".
51. Bhatnagar, "Yes and No".

Chapter 5: Spirits

1. See the analysis in Dove, "Dialectical History of 'Jungle'".
2. Omvedt, *Dalits and Democratic Revolution*, 36–7.
3. Bobb and Gupta, *Delhi Then and Now*, 13.
4. An excellent analysis of the Nath siddhis, in both their historical and contemporary manifestations, is White, *The Alchemical Body*.
5. For the synergies between these three professions, see Gommans, "The silent frontier".
6. This is reflected in folk tales like the *Alha*, analyzed later in this chapter. See Hiltebeitel, *Rethinking India's Epics*.
7. White, *The Alchemical Body*, 8.
8. More details on this surprising syncretism can be found in Bouillier, "Nāth Yogīs".
9. My analysis of the *Alha* draws heavily from Hiltebeitel, *Rethinking India's Epics*.
10. Hiltebeitel, *Rethinking India's Epics*, 300.
11. Quoted in Hiltebeitel, *Rethinking India's Epics*, 168.
12. Hasan's description of this tomb is quoted in Lewis and Lewis, *Mehrauli*, 5.
13. See Hiltebeitel, *Rethinking India's Epics*, 158.

14. This fading of the Khwaja Khizr myth is recorded in Dalrymple, *City of Djinns*, 298–310.
15. Khan, "Portrait of a City", 176–7.
16. Quoted in White, *The Alchemical Body*, 412.
17. Quoted in White, *The Alchemical Body*, 349.
18. Quoted in White, *The Alchemical Body*, 349.
19. The relevant pages are van Buitenen, *The Mahabharata Book 1*, 401–3.
20. This balance is described in Baviskar, "Urban Jungles", 51.
21. See IANS, "Couple Attempt Suicide" and TNN, "Lovers found hanging".
22. Seabrook, *Love in a Different Climate*.
23. Seabrook, *Love in a Different Climate*, 172.
24. Seabrook, *Love in a Different Climate*, 52.
25. See, for instance, the account in Miller, *Delhi*, 137–8.
26. For early reporting on this, see Kaufman, "Riches Gone"; a later summary of reporting can be found in Shekhar, "Lonely in life".
27. Barry, "The Jungle Prince".
28. This point is emphasized in Baviskar, "Urban Jungles", 50–1.
29. The details of the crime can be found in the court case, State vs. Jasbir Singh.
30. Sealy, "First In, Last Out", 65–83.
31. Sealy, "First in, Last Out", 80.
32. This account is from a Hindi newspaper quoted in Nigam, "Theatre of the Urban", 23.
33. See Nigam, "Theatre of the Urban", 24.
34. For a similar argument, see Sethi, "The Monkeyman of Delhi".
35. See Nigam, "Theatre of the Urban", 27.
36. For Adityanath's rise, see Gatade, "Hindutvaisation of Gorakhnath Mutt"; for an analysis of the syncretism he is betraying, see Karwoski, "Far from Hindutva".
37. My account of Israil Camp draws from the expertly reported story by Neha Dixit, "A shanty town".
38. Dixit, "A shanty town".
39. Quoted in Dixit, "A shanty town".
40. Quoted in Dixit, "A shanty town".
41. Ashraf, "Slum cluster Vasant Kunj".
42. For an early assessment of AAP's "post-ideological" posturing, see Crowley, "India's Post-Ideological Politician"; for a later assessment, see Kumar, "Requiem For AAP".
43. Press Trust of India, "Aam Aadmi Party"; Press Trust of India, "Ravidas Temple Issue".
44. For more on this dubious enzyme, see Verchot, "Happy Birthday Sri Sri".
45. Tripathi, *Dera Sacha Sauda*.

46. Ashis Nandy has convincingly argued that the spate of new godmen (and godwomen) are actually functional replacements for *isht devta*, or personal gods. With mass migration, breakneck urbanization and the attending social upheaval, these gods are disappearing, and people are looking for guidance that is at once practical and spiritual. Singh and his ilk fill this role for the unsettled, uncertain, aspirational masses. See Nandy, "Indians love Radhe Maa".

47. For more on this, see Sambhav, "Indigenous Civil Engineers".

48. See, for instance, Hafeez, "Wazirpur steel factory workers".

BIBLIOGRAPHY

Primary Sources

In this section of the bibliography, I have included archival sources, government reports, court cases, pamphlets and booklets from civil society organizations, environmental assessment reports and newspaper articles.

Ahlawat, Bijendra. "Forest official's kin buys land in Aravallis, probe ordered." *The Tribune*, January 15, 2020. https://www.tribuneindia.com/news/forest-official%E2%80%99s-kin-buys-land-in-aravallis-probe-ordered-26898

Ambience Developers Pvt. Ltd. "Environmental Impact Assessment, Ambi Mall—Vasant Kunj, Delhi." New Delhi, March 2006.

Ashok, Sowmiya. "Asola Sanctuary encroachments razed." *The Hindu*, April 29, 2014. http://www.thehindu.com/news/cities/Delhi/asola-sanctuary-encroachments-razed/article5958938.ece

Ashraf, Ajaz. "In a slum cluster in Delhi's posh Vasant Kunj, the verdict is clear: AAP is headed for victory." *Scroll.in*, January 30, 2015. https://scroll.in/article/703162/in-a-slum-cluster-in-delhis-posh-vasant-kunj-the-verdict-is-clear-aap-is-headed-for-victory

Barry, Ellen."The Jungle Prince of Delhi." *New York Times*, Nov 22, 2019. https://www.nytimes.com/2019/11/22/world/asia/the-jungle-prince-of-delhi.html

Bhatnagar, Gaurav Vivek. "Yes and No over Bhatti Mines." *The Hindu*, March 24, 2013. http://www.thehindu.com/news/cities/Delhi/yes-and-no-over-bhatti-mines/article4543939.ece

Chauhan, Neeraj and Dwaipayan Ghosh. "Ponty Chadha killed in shootout with brother." *The Times of India*, November 18, 2012. https://timesofindia.indiatimes.com/india/Ponty-Chadha-killed-in-shootout-with-brother/articleshow/17260484.cms

Coventry, B.O. *Scheme for the Afforestation of the Ridge, Delhi.* Delhi: Superintendent Government Printing, 1913.

Delhi Development Authority. *Master Plan for Delhi.* Delhi: DDA, 1962.

Delhi Development Authority. *Master Plan for Delhi, 2001.* New Delhi: DDA, August 1, 1990.

Delhi Planning Department. "Economic Survey of Delhi 2012–3." Government of NCT of Delhi, 2014. http://delhi.gov.in/wps/wcm/connect/DoIT_Planning/planning/misc./economic+survey+of+delhi+2012–13

Delhi Town Planning Committee. *First report of the Delhi town planning Committee on the choice of a site for the new imperial capital with Two maps*. Her Majesty's Stationery Office, London, 1913.

Department of Forests and Wildlife. "Pending Court Cases." Government of the NCT of Delhi, November 30, 2008. http://www.delhi.gov.in/wps/wcm/connect/965f2e804f26da9c85d2bfb9b14a7e32/Panding_Court_Cases.xls?MOD=AJPERES&CACHEID=965f2e804f26da9c85d2bfb9b14a7e32

Deputy Commissioner's Office [D.C.O]. Delhi State Archives [D.S.A].

Dixit, Neha. "A shanty town in South Delhi pays a high price for resisting communal polarisation." *Scroll.in*, November 28, 2014. https://scroll.in/article/692011/a-shanty-town-in-south-delhi-pays-a-high-price-for-resisting-communal-polarisation

Dogra, Chander Suta. "Aravallis being gobbled up by land developers." *The Hindu*, February 8, 2013.

Dogra, Chander Suta. "How village common property along the Aravallis is grabbed." *The Hindu*, February 9, 2013. https://www.thehindu.com/news/how-village-common-property-along-the-aravallis-is-grabbed/article4394445.ece#:~:text=Rozka%20Gujjar%20is%20an%20uninhabited,on%20the%20outskirts%20of%20Faridabad.&text=Land%20of%20this%20village%20has,a%20suburban%20township%20of%20Gurgaon).

EST Consultants Pvt. Ltd. "Rapid Environmental Impact Assessment & Environmental Management Plan, 'Emporio' Shopping Complex, Plot No. 4, Vasant Kunj, Delhi, Executive Summary." New Delhi, December 2005.

EST Consultants Pvt. Ltd. "Rapid Environmental Impact Assessment & Environmental Management Plan, 'Promenade' Shopping Complex, Plot No. 3, Vasant Kunj, Delhi, Executive Summary." New Delhi, December 2005.

Express News Service. "Retailers down shutters, protest high rent at DLF malls." *The Indian Express*. April 11, 2009. http://archive.indianexpress.com/news/retailers-down-shutters-protest-high-rent-at-dlf-malls/445688/

Gandhiok, Jasjeev. "Reclaimed land on Southern Ridge to get identity tag." *Times of India*, Aug 21, 2019. https://timesofindia.indiatimes.com/city/delhi/reclaimed-land-on-southern-ridge-to-get-identity-tag/articleshow/70777735.cms

Government of India. "Report of Commission of Inquiry on Maruti Affairs, 1977." New Delhi: Government of India, 1997.

Government of NCT of Delhi. "Report of the Committee to Recommend the Pattern of Management of the Delhi Ridge." Delhi, 1993.

Hafeez, Sarah. "Wazirpur steel factory workers continue strike, demand rise in wages." *The Indian Express*, June 27, 2014. http://indianexpress.com/article/cities/delhi/wazirpur-steel-factory-workers-continue-strike-demand-rise-in-wages/

IANS. "Couple Attempt Suicide in Delhi." *Yahoo! News*. February 4, 2013. https://in.news.yahoo.com/couple-attempt-suicide-delhi-124407455.html

IANS. "K.P. Singh almost severed links with DLF." *Hindustan Times*, November 15, 2011. https://www.hindustantimes.com/books/k-p-singh-almost-severed-links-with-dlf/story-y6O2yZnpVlm9EYi86e7NQP.html

Jasmine Projects Pvt. Ltd. "Executive Summary." New Delhi, March 2005.

Jenkins, Cordelia, and Aman Malik. "Deepak Bhardwaj: Steady rise, sudden death." *Livemint*, April 28, 2013. http://www.livemint.com/Specials/bwyIvT5Oe7Ms55tYbbbbtI/Steady-rise-sudden-death.html

Kalpavriksh. *The Delhi Ridge Forest: Decline and Conservation*. New Delhi: Kalpavriksh, 1991.

Kannoth, Surya R., and Mandar Nimkar. "Realty faces reality: 15–20% correction by Q1 2009." *The Economic Times*, November 1, 2008.

Kaufman, Michael. "Riches Gone, India Princess Reigns in Rail Station." *New York Times*, Sept 9, 1982.

Kumar, Ashok. "Birdwatchers Thrashed at Mangar Forest." *The Hindu*, March 31, 2014. http://www.thehindu.com/news/cities/Delhi/birdwatchers-thrashed-at-mangar-forest/article5853663.ece

Kumar, K.P. Narayana. "Gurgaon on its deathbed: Haphazard model of development causes severe water crisis." *The Economic Times*, February 2, 2014. http://economictimes.indiatimes.com/news/economy/infrastructure/gurgaon-on-its-deathbed-haphazard-model-of-development-causes-severe-water-crisis/articleshow/29728053.cms

Monga, Neeraj, and Nitin Mangal. *DLF Limited: A Crumbling Edifice*. Ontario: Veritas Investment Research, 2012.

Nandi, Jayashree. "Now, revenue dept to demarcate Ridge area." *Times of India*, September 23, 2014. http://timesofindia.indiatimes.com/city/delhi/Now-revenue-dept-to-demarcate-Ridge-area/articleshow/43187840.cms

Narayan, Shalini, and Apurva. "The Making of a Tycoon." *Indian Express*, April 14, 2013. http://archive.indianexpress.com/news/the-making-of-a-tycoon/1102056/0

National Archives of India [N.A.I.]. *From Ghalib's Dilli to Lutyens' New Delhi: a Documentary Record*, edited by Mushiral Hasan and Dinyar Patel. Selections from the N.A.I. reprinted. New Delhi: Oxford University Press, 2014.

National Capital Region Planning Board. *Delhi 1999: a fact sheet*. Delhi, 1999.

Panditi, Ambika. "Signature Bridge hits a Rocky Patch." *Times of India*, September 14, 2014. http://timesofindia.indiatimes.com/city/delhi/Signature-Bridge-hits-a-rocky-patch/articleshow/42579410.cms

People's Union for Democratic Rights. *Driving Force: Labour Struggles and Violation of Rights in Maruti Suzuki India Limited*. Delhi: PUDR, 2013.

Press Trust of India. "Aam Aadmi Party To Raise Railways Demolition Drive Issue in Parliament." *NDTV*, December 14, 2015. https://www.ndtv.com/india-news/aam-aadmi-party-to-raise-railways-demolition-drive-issue-in-parliament-1254635

Press Trust of India. "Deepak Bhardwaj murder case cracked, son arrested." *Indian Express*, April 9, 2013. http://archive.indianexpress.com/news/deepak-bhardwaj-murder-case-cracked-son-arrested/1099795/0

Press Trust of India. "Delhi ridge: NGT castigates AAP govt, slaps Rs 2L fine." *DNA India*, July 20, 2017. https://www.dnaindia.com/india/report-delhi-ridge-ngt-castigates-aap-govt-slaps-rs-2l-fine-2508690

Press Trust of India. "Developers of Ambience Mall violated conditions: NGT." *Business Standard*, October 24, 2013. https://www.business-standard.com/article/pti-stories/developers-of-ambience-mall-violated-conditions-ngt-113102401128_1.html

Press Trust of India. "NGT prohibits dumping of rubble in ridge area." *ZeeNews*, May 15, 2013. https://zeenews.india.com/news/eco-news/ngt-prohibits-dumping-of-rubble-in-ridge-area_848826.html

Press Trust of India. "Ravidas Temple Issue: AAP Govt Asks DDA to Start Process of Land Denotification." *TheWire.in*, August 29, 2019. https://thewire.in/government/ravidas-temple-issue-aap-govt-asks-dda-to-start-process-of-land-denotification

Press Trust of India. "Spiritual traditions a force that has run India: PM Narendra Modi." *Times of India*, March 4, 2019. https://timesofindia.indiatimes.com/india/spiritual-traditions-a-force-that-has-run-india-pm-narendra-modi/articleshow/68260075.cms

Punjab Government. *Gazetteer of Delhi, 1883–4*. Reprinted from 1884 original. Gurgaon: Vintage Books, 1988.

Punjab Government. *Gazetteer of the Delhi District, 1912.* Lahore, 1912.

Rao, V. Venkateswara. "Black, bold and bountiful." *The Hindu Business Line,* August 13, 2010.

Roy, Dunu. *Pollution, Pushta, and Prejudices.* Delhi: Hazards Centre, 2004.

Shekhar, Raj. "Lonely in life, this Avadh 'prince' died a pauper." *Times of India,* Nov 7, 2017. http://timesofindia.indiatimes.com/articleshow/61537207.cms

Singh, Darpan. "Delhi: Forest dept suspects 'foreign' nationals living illegally in Asola sanctuary." *Hindustan Times,* March 20, 2014. http://www.hindustan times.com/india/delhi-forest-dept-suspects-foreign-nationals-living-illegally-in-asola-sanctuary/story-a2IVU7SwjIlDqFRZmBbVTM.html

Singh, Shalini. "Behind Robert Vadra's fortune, a maze of questions." *The Hindu,* October 18, 2016. http://www.thehindu.com/news/national/Behind-Robert-Vadra's-fortune-a-maze-of-questions/article12550025.ece

Sinha, G.N., ed. *An Introduction to the Delhi Ridge.* New Delhi: Department of Forests & Wildlife, Govt. of NCT of Delhi, 2014.

Srishti and WWF-India. *Saving the Delhi Ridge: One Year of Conservation Action.* Srishti and WWF-India: Delhi, 1994.

Srivastava, Samar. "DLF: The ailing giant." *Forbes India,* November 16, 2015. http://www.forbesindia.com/article/india-rich-list-2015/dlf-the-ailing-giant/41357/1

Staff Reporter. "Residents term public hearings by pollution control panel a 'farce': Issue of construction of malls and office complexes near Vasant Kunj." *The Hindu,* July 5, 2006.

State vs. Jasbir Singh @ Billa and Kuljeet Singh @ Ranga. 17 (1980) DLT 404, ILR 1979 Delhi 571. https://indiankanoon.org/doc/1359524/

TNN. "Lovers found hanging inside park." *The Times of India,* May 4, 2014. http://timesofindia.indiatimes.com/city/delhi/Lovers-found-hanging-inside-park/articleshow/34649787.cms

Ultra Tech. "Executive Summary of Rapid Environmental Assessment report for 'ONGC Videsh Limited Office Complex.'" New Delhi, 2005.

Verma, Sanjeev. "Haryana TCP dept comes under HC scanner." *Hindustan Times,* October 7, 2013. https://www.hindustantimes.com/punjab/haryana-tcp-dept-comes-under-hc-scanner/story-icF9dTWKCMGFmAQQtxA36N.html

World Health Organization. "WHO Global Urban Ambient Air Pollution Database (update 2016)." Accessed March 9, 2020. http://www.who.int/phe/health_topics/outdoorair/databases/cities/en/

Secondary Sources

In this section, I have included scholarly works, as well as literature, memoirs, op-eds and other published material.

Agarwal, Ravi. "Fight for a Forest." *Seminar Magazine*, September 2010. http://www.india-seminar.com/2010/613/613_ravi_agarwal.htm

Anderson, Perry. "Why Partition." *London Review of Books* 34, no. 14 (2012): 1–19.

Badger, Emily. "The Shopping Mall Turns 60 (and Prepares to Retire)." *CityLab*, July 13, 2012. http://www.citylab.com/design/2012/07/shopping-mall-turns-60-and-prepares-retire/2568/

Bajaj, Sheela. "The Building of New Delhi." In *City Improbable: An Anthology of Writings on Delhi*, edited by Khushwant Singh, 63–7. New York: Viking, 2001.

Bandyopadhyay, Sekhar. *From Plassey to Partition: A History of Modern India*. Hyderabad: Orient Blackswan, 2004.

Baviskar, Amita. "Between violence and desire: space, power, and identity in the making of metropolitan Delhi." *International Social Science Journal* 55, no. 175 (2003): 89–98.

Baviskar, Amita. "Cows, Cars and Cycle-rickshaws: Bourgeois Environmentalists and the Battle for Delhi's Streets." In *Elite and Everyman: the Cultural Politics of the Indian Middle Classes*, edited by Amita Baviskar and Raka Ray, 391–418. Delhi: Routledge, 2011.

Baviskar, Amita, ed. *First Garden of the Republic: Nature on the President's Estate*. New Delhi: Publications Division, Government of India, 2016.

Baviskar, Amita. "Spectacular Events, City Spaces and Citizenship: The Commonwealth Games in Delhi." In *Urban Navigations: Politics, Space and the City in South Asia*, edited by Jonathan Shapiro Anjaria and Colin McFarlane, 152–176. New Delhi: Routledge, 2011.

Baviskar, Amita. "Urban Jungles: Wilderness, Parks and Their Publics in Delhi." *Economic and Political Weekly* 53, no. 2 (2018): 46–54.

Bayley, Emily, and Thomas Metcalfe. *Golden Calm: An English Lady's Life in Moghul Delhi: Reminiscences*. London: Penguin Books, 1980.

Beg, Mirza Farhatullah. *Bahadur Shah and the Festival of Flower-Sellers*. Translated by Mohammed Zakir. Hyderabad: Orient Blackswan, 2012.

Bhan, Gautam. "Planned Illegalities: Housing and the 'Failure' of Planning in Delhi: 1947–2010." *Economic & Political Weekly* 48, no. 24 (2013), 58–70.

Blake, Stephen. *Shahjahanabad: the Sovereign City in Mughal India, 1639–1739*. Cambridge: Cambridge University Press, 1991.

Bobb, Dilip, and Narayani Gupta. *Delhi Then and Now*. Delhi: Roli Books, 2008.

Botkin, Daniel. *Discord Harmonies: A New Ecology for the 21st Century*. Oxford: Oxford University Press, 1992.

Bouillier, Véronique. "Nāth Yogīs' Encounters with Islam." *South Asia Multidisciplinary Academic Journal*, 2015. http://samaj.revues.org/3878

Byrnes, Mark. "Victor Gruen Wanted to Make Our Suburbs More Urban. Instead, He Invented the Mall." *CityLab*, July 18, 2013. http://www.citylab.com/design/2013/07/victor-gruen-wanted-make-our-suburbs-better-instead-he-invented-mall/6249/

Chakravarty-Kaul, Minoti. *Common Lands and Customary Law: Institutional Change in North India Over the Past Two Centuries*. Delhi: Oxford University Press, 1996.

Chandra, Satish. *Essays on Medieval Indian History*. Delhi: Oxford University Press, 2006.

Chattopadhyaya, B.D. "Emergence of the Rajputs as Historical Process in Early Medieval Rajasthan." In *The Idea of Rajasthan: Explorations in Regional Identity, Volume II: Institutions*, edited by Karine Shomer, J.L. Erdman and D.O. Lodrick, 161–191. Delhi: South Asia Publications, 1994.

Chibber, Vivek. *Locked in Place: State Building And Late Industrialization In India*. Delhi: Tulika Books, 2004.

Christ, Kilian Alexander. "Interaction between water related informal processes and water management: the example of Hyderabad, India." PhD dissertation, RWTH Aachen University, 2015.

Crowley, Thomas. "Demolitions in Aya Nagar, Delhi." *Kafila*, September 13, 2014. https://kafila.online/2014/09/13/demolitions-in-aya-nagar-delhi-thomas-crowley/

Crowley, Thomas. "India's Post-Ideological Politician." *Jacobin*, December 11, 2013. https://www.jacobinmag.com/2013/12/indias-post-ideological-politician

Dalrymple, William. *City of Djinns: A Year in Delhi*. Delhi: Penguin India, 2004.

Dalrymple, William. *The Last Mughal: The Fall of a Dynasty: Delhi, 1857*. New York: Vintage, 2008.

Diamond, Jared. "The Worst Mistake in the History of the Human Race." *Discover Magazine*. May 1, 1999. http://discovermagazine.com/1987/may/02-the-worst-mistake-in-the-history-of-the-human-race

Dirks, Nicholas. *The Scandal of Empire: India and the Creation of Imperial Britain*. Cambridge: Harvard University Press, 2009.

Donthi, Praveen. "Under Jagmohan, Jammu and Kashmir entered a period of unfettered repression." *The Caravan*, January 25, 2016. http://www.caravanmagazine.in/vantage/jagmohan-padma-vibhushan-kashmir

Dove, Michael. "The Dialectical History of 'Jungle' in Pakistan: An Examination of the Relationship between Nature and Culture." *Journal of Anthropological Research* 48, no. 3 (1992): 231–53.

Fanshawe, Herbert Charles. *Delhi: Past and Present*. London: John Murray, 1902.

Gatade, Subhash. "Hindutvaisation of a Gorakhnath Mutt: the Yogi and the Fanatic." *South Asia Citizens Web*, October 7, 2004. http://www.sacw.net/DC/CommunalismCollection/ArticlesArchive/gatade07102004.html

Ghertner, Asher. "Green evictions: environmental discourses of a 'slum-free' Delhi." In *Global Political Ecology*, edited by Richard Peet, Paul Robbins, and Michael Watts, 159–180. New York: Routledge, 2011.

Goel, Sita Ram. *Heroic Hindu Resistance to Muslim Invaders (636 AD to 1206 AD)*. Meerut: Anu Books, 1983.

Gommans, Jos. "The silent frontier of South Asia, C. A.D. 1100–1800." *Journal of World History* 9, no 1 (1998): 1–23.

Gooptu, Nandini. "Neoliberal Subjectivity, Enterprise Culture and New Workplaces: Organised Retail and Shopping Malls in India." *Economic & Political Weekly* 44, no. 22 (2009): 45–54.

Graebar, David, and David Wengrow. "How to change the course of human history (at least, the part that's already happened)." *Eurozine*, March 2, 2018. https://www.eurozine.com/change-course-human-history/

Guha, Ranajit. *A Rule of Property for Bengal: An Essay on the Idea of Permanent Settlement*. Durham: Duke University Press, 1996.

Gupta, Narayani. "Delhi and its Hinterlands: The 19th and Early 20th Centuries." In *Delhi Through the Ages: Essays in Urban History, Culture, and Society*, edited by R.E. Frykenberg, 250–69. Oxford: Oxford University Press, 1986.

Hansen, Peter H. "Albert Smith, the Alpine Club, and the Invention of Mountaineering in Mid-Victorian Britain." *Journal of British Studies* 34, no. 3 (1995): 300–324. doi:10.1086/386080.

Hashmi, Sohail. "The Lives And Times Of The Asokan Pillar At The Delhi Ridge." *Kafila*, October 31, 2007. https://kafila.online/2007/10/31/the-lives-and-times-of-the-asokan-pillar-at-the-ridge/

Hiltebeitel, Alf. *Rethinking India's Oral and Classical Epics: Draupadi among Rajputs, Muslims and Dalits.* Chicago: University of Chicago Press, 1999.

Jagmohan. *The Challenge of Our Cities.* Delhi: Vikas Publishing House, 1984.

Jagmohan. *Island of Truth.* Delhi: Vikas Publishing House, 1978.

Jagmohan. *My Frozen Turbulence.* Delhi: Allied Publishers, 1991.

Jagmohan. *Shaping India's New Destiny.* Delhi: Allied Publishers: 2008.

Karve, Iravate. *Yuganta: the End of an Epoch.* Hyderabad: Orient Blackswan, 2016.

Karwoski, C. Marrewa. "Far from Hindutva, Yogi Adityanath's sect comes from a tradition that was neither Hindu nor Muslim." *Scroll.in*, April 9, 2017. https://scroll.in/article/833710/far-from-hindutva-yogi-adityanath-comes-from-a-tradition-that-was-neither-hindu-nor-muslim

Khan, Dargah Quli. "Portrait of a City." Translated by Saleem Kidwai. In *Same-Sex Love in India*, edited by Ruth Vanita and Saleem Kidwai, 175–83. New York: St. Martin's Press, 2000.

Khari, Rahul. *Jats and Gujars: Origin, History and Culture.* Delhi: Reference Press, 2007.

Kohn, Marek, and Steven Mithen. "Handaxes: Products of sexual selection?" *Antiquity* 73 (1999): 518–26.

Krishen, Pradip. *Trees of Delhi: A Field Guide.* Delhi: Dorling Kindersley, 2006.

Kumar, Ravi. "Punjab, Goa And Delhi MCD Elections: Scripting A Requiem For AAP." *Huffpost*, May 2, 2017. https://www.huffingtonpost.in/ravi-kumar/punjab-goa-and-delhi-mcd-elections-scripting-a-requiem-for-aap_a_22063227

Kumar, Sunil. *The Emergence of the Delhi Sultanate, 1192–1286.* Ranikhet: Permanent Black, 2007.

Kumar, Sunil. *The Present in Delhi's Pasts: Five Essays.* Gurgaon: Three Essays Collective, 2010.

Lahiri, Nayanjot. *Ashoka in Ancient India.* Cambridge: Harvard University Press, 2015.

Lahiri, Nayanjot, and D.K. Chakrabarti. "A preliminary report on the stone age of the union territory of Delhi and Haryana." *Man and Environment* 11 (1987): 109–16.

Lam, Mimi. "The First Commodity: Handaxes." Paper presented at the American Association for the Advancement of Science 2012 Annual Meeting, February 2012.

Lewis, Karoki, and Charles Lewis. *Mehrauli: a view from the Qutb.* Delhi: HarperCollins, 2002.

Maestripieri, Dario. *Macachiavellian Intelligence: how Rhesus Macaques and Humans have Conquered the World.* Chicago: University of Chicago Press, 2007.

Mann, Michael. "Delhi's Belly: On the Management of Water, Sewage and Excreta in a Changing Urban Environment during the Nineteenth Century." *Studies in History* 23, no 1 (2007), 1–31.

Mann, Michael, and Samiksha Sehrawat. "City with a View: the Afforestation of the Delhi Ridge, 1883–1913." *Modern Asian Studies* 43, no. 2 (2009): 543–70.

Marx, Karl. *Capital: Volume 3.* New York: International Publishers, 1967.

Marx, Karl. "Economic and Philosophical Manuscripts of 1844." In *The Marx-Engels Reader*, edited by Robert C. Tucker, 66–125. New York: W.W. Norton & Company, 1978.

Mawdsley, Emma. "India's Middle Classes and the Environment." *Development and Change* 35, no. 1 (2004): 79–103.

Mayaram, Shail. "Pastoral Predicaments: The Gujars in history." *Contributions to Indian Sociology* 48 (2014): 191–222.

Mehra, Diya. "Planning Delhi ca. 1936–1959." *South Asia: Journal of South Asian Studies* 36, no. 3 (2013): 354–74. DOI: 10.1080/00856401.2013.829793.

Mellars, Paul, Kevin C. Goric, Martin Carre, Pedro A. Soares, and Martin B. Richards. "Genetic and archaeological perspectives on the initial modern human colonization of southern Asia." *Proceedings of the National Academy of Sciences of the United States of America* 110, no. 26 (2013): 10699–10704.

Miller, Sam. *Delhi: Adventures in a Megacity.* New York: St. Martin's Press, 2009.

Mishra, Pankaj. "Modi's Idea of India." *The New York Times*, October 24, 2014. https://www.nytimes.com/2014/10/25/opinion/pankaj-mishra-nirandra-modis-idea-of-india.html

Morrison, Kathleen. "Conceiving Ecology and Stopping the Clock: Narratives of Balance, Loss and Degradation." In *Shifting Ground: People, Animals and Mobility in Indian Environmental History*, edited by Mahesh Rangarajan and Kalyanakrishnan Sivaramakrishnan, 39–64. Oxford: Oxford University Press, 2014.

Nandy, Ashis. "There's a reason Indians love Radhe Maa. And it's not superstition." *Catch News*, August 15, 2015. http://www.catchnews.com/india-news/there-s-a-reason-indians-love-radhe-maa-and-it-s-not-superstition-1439445272.html

Negi, Rohit. "Neoliberalism, Environmentalism and Urban Politics in Delhi." in *India's New Economic Policy: A Critical Analysis*, edited by Waquar Ahmed, Amitabh Kundu and Richard Peet, 179–98. London: Routledge, 2011.

Nigam, Aditya. "Industrial Closures in Delhi." *Revolutionary Democracy* 8, no. 2 (2001). http://www.revolutionarydemocracy.org/rdv7n2/industclos.htm

Nigam, Aditya. "Theatre of the Urban: the Strange Case of the Monkeyman." In *Sarai Reader 02: The Cities of Everyday Life*, edited by Ravi S. Vasudevan, Jeebesh Bagchi, Ravi Sundaram, Monica Narula, Geert Lovink & Shuddhabrata Sengupta, 22–30. Delhi/Amsterdam: Sarai + Society for Old and New Media, 2002.

Omvedt, Gail. *Dalits and the Democratic Revolution: Dr. Ambedkar and the Dalit Movement in Colonial India*. Delhi: Sage, 1994.

Parashar, Arpit. "The natives strike back at the Raj." *Foundation Ink*, July 7, 2012. http://fountainink.in/?p=2274&all=1

Parashar, Arpit. "Still waiting for a Raisina welcome." *Foundation Ink*, March 5, 2015. https://fountainink.in/reportage/still-waiting-for-a-raisina-welcome

Parker, R.N. "Afforestation of the Ridge at Delhi." *The Indian Forester* 46, no. 1 (1920): 22–28.

Pati, Sushmita. "Jagmohan: The Master Planner and the 'Rebuilding' of Delhi." *Economic & Political Weekly* 49, no. 36 (2014): 48–54.

Petraglia, Michael, Christopher Clarkson, Nicole Boivin, Michael Haslam, Ravi Korisettar, Gyaneshwer Chaubey, Peter Ditchfield et al. "Population increase and environmental deterioration correspond with microlithic innovations in South Asia ca. 35,000 years ago." *Proceedings of the National Academy of Sciences of the United States of America* 106, no. 30 (2009): 12261–12266.

Ratnam, Jayashree, William J. Bond, Rod J. Fensham, William mann, Sally Archibald, Caroline E. R. Lehmann, Michael T. Anderson   hen is a 'forest' a savanna, and why does it matter?" *Global Ecology an   eography* 20, no. 5 (2011): 653–60. DOI: 10.1111/j.1466-8238.2010.00634.x.

Richard, A.F., S.J. Goldstein and R.E. Dewar. "Weed macaques: The evolutionary implications of macaque feeding ecology." *International Journal of Primatology* 10 (1989): 569–91. doi: 10.1007/BF02739365

Rosler, Martha. "Culture Class: Art, Creativity, Urbanism, Part III." *e-flux* 25 (2011). http://www.e-flux.com/journal/25/67898/culture-class-art-creativity-urbanism-part-iii/

Sambhav, Kumar. "Indigenous Civil Engineers." *Down to Earth*, February 15, 2015. http://www.downtoearth.org.in/content/indigenous-civil-engineers

Scott, James. *Against the Grain: a Deep History of the Earliest States*. New Haven: Yale University Press, 2017.

Seabrook, Jeremy. *Love in a Different Climate: Men Who Have Sex with Men in India*. London: Verso, 1999.

Sealy, Irwin Allen. "First In, Last Out." In *Delhi Noir*, edited by Hirsh Sawney, 65–83. New York: Akashic Books, 2009.

Searle, Llerena Guiu. "Conflict and Commensuration: Contested Market Making in India's Private Real Estate Development Sector." *International Journal of Urban and Regional Research* 38, no. 1 (2014), 60–78.

Searle, Llerena Guiu. "The contradictions of mediation: intermediaries and the financialization of urban production." *Economy and Society* 47, no. 4 (2018), 524–46.

Sengupta, Ranjana. *Delhi Metropolitan: The Making of an Unlikely City*. New Delhi: Penguin Books, 2008.

Sethi, Aman. "The Monkeyman of Delhi." *Guernica*, October 1, 2012. https://www.guernicamag.com/the-monkeyman-of-delhi/

Sharma, A.K. "Prehistoric Delhi and its Neighborhood: Physical Features." In *Delhi: Ancient History*, edited by Upinder Singh, 14–25. New Delhi: Social Science Press, 2006.

Shrivastava, Kumar Sambhav. "Simians' Sanctuary." *Down To Earth*, February 15, 2015. http://www.downtoearth.org.in/coverage/indigenous-civil-engineers-48425

Singh, Malvika, Rudrangshu Mukherjee, and Pramod Kapoor. *New Delhi: Making of a Capital*. Delhi: Lustre Press, 2009.

Singh, Upinder. *Ancient Delhi*. Delhi: Oxford University Press, 2010.

Singh, Upinder. "The Later Histories of the Ashokan and Mehrauli Pillars." In *Delhi: Ancient History*, edited by Upinder Singh, 207–11. New Delhi: Social Science Press, 2006.

Smith, Neil. "Nature as Accumulation Strategy." *Socialist Register* 43 (2007): 16–36.

Soni, Anita. "Displacement Woes I: Tell Us Where to Go." *Tehelka*, July 2006.

Soni, Anita. "Displacement Woes II: Use us, Don't Abuse Us Please." *Tehelka*, July 2006.

Soni, Anita. "Urban Conquest of Outer Delhi: Beneficiaries, Intermediaries and Victims." In *Delhi: Urban Space and Human Destinies*, edited by Veronique Dupont, Emma Tarlo and Denis Vidal, 69–91. Delhi: Manohar, 2000.

Soni, Vikram. *Naturally: Tread Softly on the Planet*. Delhi: HarperCollins India, 2015. Kindle.

Srivastava, Sanjay. *Entangled Urbanism: Slum, Gated Community, and Shopping Mall in Delhi and Gurgaon*. Delhi: Oxford University Press, 2014.

Stanley, Peter. *White Mutiny: British Military Culture in India*. New York: New York University Press, 1998.

Stokes, Eric. *The Peasant and the Raj: Studies in Agrarian Society and Peasant Rebellion in Colonial India*. Cambridge: Cambridge University Press, 1978.

Sundaram, Ravi. *Pirate Modernity: Delhi's Media Urbanism*. London: Routledge, 2010.

Talib, Mohammad.*Writing Labour: Stone Quarry Workers in Delhi*. Delhi: Oxford University Press, 2010.

Thapar, Romila. *A History of India, Vol. One*. Baltimore: Penguin Books, 1966.

Timur. *Malfūzāt-e Tīmūrī* [excerpts]. In *History of India,Vol V: The Mohammedan Period as Described by its Own Historians*, edited by A.V. Williams Jackson. 1907. Reprint of the first edition. New York: Cosimo Classics, 2008.

Tripathi, Anurag. *Dera Sacha Sauda and Gurmeet Ram Rahim: A Decade-long Investigation*. Gurugram: Penguin Random House India, 2018.

Tripathi, R.S. *History of Kanauj*. Delhi: Motilal Banarsidass, 1964.

Trivedi, Mudit. "On the Surface Things Appear to be... Perspectives on the Archeology of the Delhi Ridge." In *Ancient India: New Research*. Edited by

Upinder Singh and Nayanjot Lahiri, 39–71. Delhi: Oxford University Press, 2010.

van Buitenen, J.A.B., trans. *The Mahabharata, Volume 1: Book 1: The Book of the Beginning*. Chicago: University of Chicago Press, 1980.

Varma, Pavan, and Sondeep Shankar. *Mansions at dusk: the havelis of old Delhi*. Delhi: Spantech Publishers, 1992.

Verchot, Manon. "Happy Birthday Sri Sri, Writing About You Has Been Bittersweet." *The Quint*, May 13, 2016. https://www.thequint.com/india/2016/05/13/sri-sri-ravi-shankar-birthday-art-of-living-yamuna-floodplain

Verma, P.K., and R.O. Greiling. "Tectonic evolution of the Aravalli orogen (NW India): an inverted Proterozoic rift basin?" *Geologische Rundschau* 84 (1995): 683–96. doi: 10.1007/BF00240560

Vij, Shivam. "The BSP Billionaires." *Open Magazine*, April 25, 2009. http://www.openthemagazine.com/article/india/the-bsp-billionaires

Vij, Shivam. "Monkey Business." *Motherland*, January 1, 2013. http://www.motherlandmagazine.com/monkey-business

Vij, Shivam. "Why the Indian Parliament's monkey problem has no easy solution." *Scroll.in*, August 1, 2014. https://scroll.in/article/672459/why-the-indian-parliaments-monkey-problem-has-no-easy-solution

Weber, Max. "Politics as a Vocation." In *The Vocation Lectures*, edited by David Owen and Tracy B. Strong, translated by Rodney Livingstone, 32–94. Indianapolis: Hackett Publishing Company, 2004.

Weisman, Alan. *The World Without Us*. New York: Picador, 2008.

White, David Gordon. *The Alchemical Body: Siddha Traditions in Medieval India*. Chicago: University of Chicago Press, 1998.

Wiedenbeck, M., J.N. Goswami, and A.B. Roy. "Stabilization of the Aravalli Craton of northwestern India at 2.5 Ga: An ion microprobe zircon study." *Chemical Geology* 129 (1996): 325–40.

Witzel, Michael. "Autochthonous Aryans? The Evidence from Old Indian and Iranian Texts." *Electronic Journal of Vedic Studies* 7, no. 3 (2001): 1–115.

Young, Keith. *Delhi—1857; the Siege, Assault, and Capture as Given in the Diary and Correspondence of the Late Colonel Keith Young*, edited by Henry Wylie Norman and Mrs. Keith Young. 1902. Reprint of the first edition. Delhi: Gian Publishing House, 1988.

ABOUT THE AUTHOR

Thomas Crowley has spent over a decade researching and writing about environmental politics and history in India. From 2010 to 2017, he conducted intensive research on the Delhi Ridge for the NGO Intercultural Resources, as well as for the "City as Studio" fellowship program at the Centre for the Study of Developing Societies (CSDS), New Delhi.

Crowley received his B.A. in Philosophy from Yale University in 2007, completed a Fulbright Scholarship at the University of Pune from 2008–2009, and served as a Social Science Fellow at the Akademie Schloss Solitude in Stuttgart, Germany from 2016–2017. He has written extensively on Indian politics for *Jacobin* magazine. He has also written for *Kafila*, as well as for peer-reviewed academic journals, including *Emotion, Space and Ecology* and *Ethics and the Environment*. He is currently researching the politics of water and caste in Maharashtra as a doctoral candidate in the Department of Geography, Rutgers University (USA).